Advanced Radar Detection Schemes Under Mismatched Signal Models

SYNTHESIS LECTURES ON SIGNAL PROCESSING

Editor
José Moura, Carnegie Mellon University

Advanced Radar Detection Schemes Under Mismatched Signal Models
Francesco Bandiera, Danilo Orlando, and Giuseppe Ricci
2009

DSP for MATLAB™ and LabVIEW™
Volume IV: LMS Adaptive Filtering
Forester W. Isen
2009

DSP for MATLAB™ and LabVIEW™
Volume III: Digital Filter Design
Forester W. Isen
2008

DSP for MATLAB™ and LabVIEW™
Volume II: Discrete Frequency Transforms
Forester W. Isen
2008

DSP for MATLAB™ and LabVIEW™
Volume I: Fundamentals of Discrete Signal Processing
Forester W. Isen
2008

The Theory of Linear Prediction
P. P. Vaidyanathan
2007

Nonlinear Source Separation
Luis B. Almeida
2006

Spectral Analysis of Signals
Yanwei Wang, Jian Li, and Petre Stoica
2006

Advanced Radar Detection Schemes Under Mismatched Signal Models
Francesco Bandiera, Danilo Orlando, and Giuseppe Ricci

ISBN: 978-3-031-01404-8 paperback
ISBN: 978-3-031-02532-7 ebook

DOI 10.1007/978-3-031-02532-7

A Publication in the Springer series
SYNTHESIS LECTURES ON SIGNAL PROCESSING

Lecture #8
Series Editor: José Moura, Carnegie Mellon University

Series ISSN
Synthesis Lectures on Signal Processing
Print 1932-1236 Electronic 1932-1694

Advanced Radar Detection Schemes Under Mismatched Signal Models

Francesco Bandiera, Danilo Orlando, and Giuseppe Ricci
Dipartimento di Ingegneria dell'Innovazione
University of Salento, Italy

SYNTHESIS LECTURES ON SIGNAL PROCESSING #8

ABSTRACT

Adaptive detection of signals embedded in correlated Gaussian noise has been an active field of research in the last decades. Such topic is important in many areas of signal processing as, just to give some examples, radar, sonar, communications, and hyperspectral imaging. Most of existing adaptive algorithms have been designed following the lead of the derivation of Kelly's detector which assumes perfect knowledge of the target steering vector. However, in realistic scenarios, mismatches are likely to occur, due to both environmental and instrumental factors. When a mismatched signal is present in the data under test, conventional algorithms may suffer severe performance degradation. The presence of strong interferers in the cell under test makes the detection task even more challenging. An effective way to cope with this scenario relies on the use of "tunable" detectors, i.e., detectors capable of changing their directivity through the tuning of proper parameters. The aim of this book is to present some recent advances in the design of tunable detectors and the focus is on the so-called two-stage detectors, i.e., adaptive algorithms obtained cascading two detectors with opposite behaviors. We derive exact closed-form expressions for the resulting probability of false alarm and the probability of detection for both matched and mismatched signals embedded in homogeneous Gaussian noise. It turns out that such solutions guarantee a wide operational range in terms of tunability, while retaining, at the same time, an overall performance in presence of matched signals commensurate with Kelly's detector.

KEYWORDS

Adaptive Radar Detection, Constant False Alarm Rate (CFAR), Interference Rejection, Mismatched Signals, Tunable Detectors

Contents

Preface

This book presents some recent developments in the area of adaptive radar detection. More precisely, it gives an in-depth coverage of the design and the analysis of tunable decision schemes as a viable means to face with detection of slightly mismatched signals buried in homogeneous Gaussian noise and possibly strong coherent interferers. It is based upon some recent work made by the authors in this context. However, in order to make the book a self-contained source for graduate students and engineers working on signal processing and its applications to radar systems, it also surveys the design of conventional (i.e., non tunable) adaptive detection schemes and contains appendices with useful mathematical derivations that can help the reader to learn the basic methodologies of analysis used throughout the entire book.

The authors acknowledge the collaboration of Olivier Besson, Antonio De Maio, and Louis L. Scharf on adaptive radar detection. They also want to express their gratefulness to Joel Claypool for publishing this work and to C. L. Tondo and M. Jones for the final editing.

Francesco Bandiera, Danilo Orlando, and Giuseppe Ricci
Dipartimento di Ingegneria dell'Innovazione
University of Salento, Italy
January 2009

CHAPTER 1

Introduction

Adaptive detection of signals buried in noise has been an important research topic for several decades [1]. Such a problem is in fact of primary concern for radar, sonar, communications, and hyperspectral imaging applications. Most of the recently proposed solutions follow the lead of the seminal paper by Kelly [2] where the generalized likelihood ratio test (GLRT) is used to conceive an adaptive decision scheme capable of detecting a coherent pulse train, embedded in Gaussian disturbance with unknown spectral properties, which impinges on an array of sensors. Kelly's detector exploits a set of training data in order to guarantee the constant false alarm rate (CFAR) property (a point better clarified in Chapter 2).

This book focuses on the design and the analysis of detection schemes for radar applications. Difficulties at the design stage come from the fact that signals of interest are often partially unknown at the receiver and from possible mismatches between the actual and the nominal steering vectors. The possible presence of (both coherent and noise like) jamming and clutter, in addition to thermal noise, makes the task of the detector even more challenging [3].

In the following we classify coherent returns into mainlobe and sidelobe signals [4, 5]

- a sidelobe signal (sidelobe interferer) is a coherent signal from a "direction significantly different" to that in which the (radar) system is steered: it can be due to a strong target located in a sidelobe direction or to the retransmission of a modulated signal (for the purpose of degrading the reception of the signal of interest);

- a mainlobe target is instead a coherent signal backscattered from the nominal direction or a direction slightly different from the nominal one as a consequence of an imperfect modeling of the nominal steering vector, where the mismatch may be due to multipath propagation, array calibration uncertainties, beampointing errors, etc. Mainlobe jamming is out of the scope of this work.

Thus, the effectiveness of the detector depends on its ability to detect the presence of what is classified as mainlobe target, limiting the number of "false alarms" due to sidelobe interferers. Ideally, the processor should guarantee a probability of detection of mainlobe targets close to one and a probability of detection of sidelobe interferers close to zero. Unfortunately, such requirements cannot be met at the same time and a viable approach is to design detectors capable of trading detection performance of mainlobe targets for rejection capabilities of sidelobe interferers.

From this point of view, decision schemes designed to detect a signal known up to a complex factor (like Kelly's detector) may behave quite differently in presence of mainlobe and sidelobe signals. Kelly's detector is rather selective in presence of mismatched signals, while the adaptive matched filter (AMF) [6] tends to be more robust. Several means have been investigated in order to increase the selectivity or the robustness of the detector. For instance, the adaptive beamformer orthogonal rejection test (ABORT) has been used to make the detector less inclined to detect signals whose steering vector is in some way orthogonal to the assumed one [4, 7]. On the other hand, an increased robustness can be achieved, as an example, by resorting to the tools of subspace detection, namely assuming that the target belongs to a known subspace of the observables [8, 9]. A way to combine the ABORT rationale with the subspace idea has been proposed in [5]. However, all of the above detectors are not tunable in the sense that they cannot easily adjust their "directivity" in order to guarantee an acceptable compromise between robustness and selectivity.

The design of tunable detectors has been pursued by different means as shown in [10] and [11, 12, 13, 14, and references therein]. The purpose of this book is to give an in-depth coverage of the design and the analysis of tunable receivers in homogeneous Gaussian environment (a point better explained in Chapter 2). The focus is on two-stage detectors that, combining (one-stage) detectors with opposite behaviors, give rise to very effective tunable decision strategies. However, the book is a self-contained source for graduate students and engineers working on signal processing. For this reason, it also surveys the design of classical (non tunable) adaptive detection schemes.

The reminder of the book is organized as follows. The next chapter introduces the discrete-time target and noise models, briefly reviews the necessary tools of statistical decision theory, and finally presents classical detection schemes as Kelly's detector, the AMF, and the adaptive coherence estimator (ACE) [15, 16], namely the ones conceived to detect signals known up to a complex factor.

Chapter 3 shows how to increase the robustness or the selectivity of a decision scheme and, eventually, how to design tunable receivers. In particular, it is shown how the tools of subspace detection have been used to come up with robust detectors; the behavior of selective detectors based on the ABORT idea and on the Rao test is also investigated. Finally, the two main classes of tunable detectors are briefly reviewed: parametric and two-stage ones; in particular, the adaptive sidelobe blanker (ASB) is introduced and its performance evaluated [17, and references therein].

Chapter 4 presents recently proposed two-stage detectors aimed at increasing either the robustness or both robustness and selectivity of the ASB [13, 14, 18]. The performance assessment is conducted analytically for both matched and mismatched signals; it is shown that such decision schemes are a viable means to attack adaptive detection in presence of uncertainties on the target steering vector and possible sidelobe interferers.

Chapter 5 summarizes presented results and briefly discusses some open issues.

The book also contains three appendices. Appendix A is a short review of the theory of the multivariate complex normal distribution and of distributions related to it, while Appendices B and C provide analytical derivations of results presented in previous chapters.

CHAPTER 2

Adaptive radar detection of targets with perfectly known steering vector

Detection of a point-like target and estimation of its relevant parameters (range, velocity, azimuth, etc.) is often complicated by signal and environmental uncertainties. In this chapter, as a preliminary step, we introduce the discrete-time models for target and noise that will be used throughout the book. Then, we review the tools used to address adaptive detection of a coherent target echo known up to a complex factor. Finally, we introduce classical algorithms for adaptive detection.

2.1 THE RADAR SCENARIO

The radar scenario involves a transmitter and a receiver, at the same location, equipped with an array of sensors, a point-like target at a certain "distance" from the array (range) in the far zone, and a narrowband signal[1] that travels the round-trip between the radar and the target.

More precisely, the system under consideration utilizes a linear array of N_a uniformly spaced and identical sensors deployed along the z axis of a preassigned reference system. The reference system is illustrated in Figure 2.1: the m-th sensor is located at $z_m = (m-1)d$, $m = 1, \dots, N_a$, d is the interelement spacing, and the spherical coordinates are R, ψ, and ϕ related to the cartesian coordinates through the equations

$$\begin{cases} x &= R \sin\psi \cos\phi, \\ y &= R \sin\psi \sin\phi, \\ z &= R \cos\psi. \end{cases}$$

In order to avoid spatial aliasing (i.e., grating lobes), it is necessary to sample in space at least twice per cycle, i.e., d has to satisfy the inequality $d \leq \lambda/2$ with λ denoting, in turn, the operating wavelength [19]. The radar transmits a coherent burst of N_p radiofrequency (RF) pulses at a constant pulse repetition frequency PRF $= 1/T$, where T is the pulse repetition time (PRT). Finally, the carrier frequency is $f_c = c/\lambda$ where c is the velocity of propagation in the medium[2].

The signal collected by the m-th sensor is amplified, filtered, and down converted: the output contains noise components[3] and possibly useful signal components. More precisely, if a target is

[1] Hereafter we assume that delay terms between sensors are insignificant within the complex envelope of the signal.
[2] Assume that $c = 3 \cdot 10^8$ m/sec.
[3] Thermal noise and, in addition, clutter and/or intentional interference.

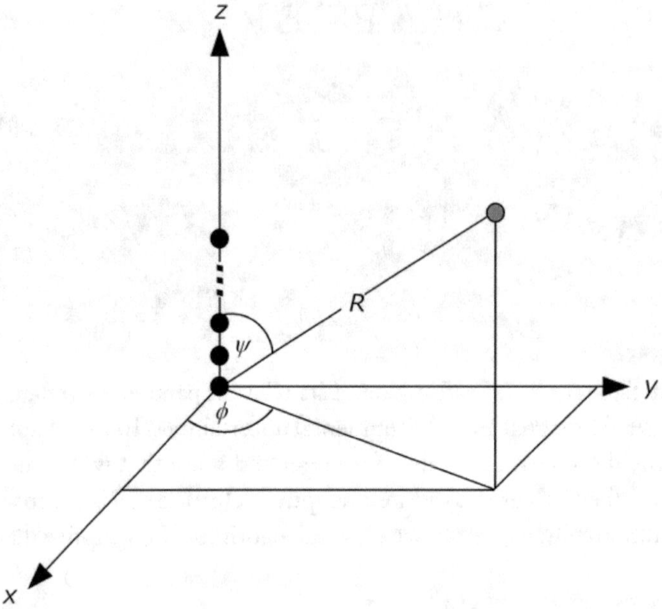

Figure 2.1: The array of sensors and the associated reference system: each sensor is represented by a filled circle.

present in the illuminated area, the complex envelope of the signal received at the m-th sensor is given by

$$r_m(t) = x_m(t) + w_m(t), \quad m = 1, \ldots, N_a, \tag{2.1}$$

where $x_m(t)$ is the (complex envelope of the) signal backscattered from the target and $w_m(t)$ is the (complex envelope of the) overall disturbance component. As to the signal backscattered from the target, we suppose that it is a delayed and attenuated copy of the transmitted one. As to the noise component, we assume the Gaussian model; more precisely, $w_m(t)$ is supposed to be a wide-sense stationary (WSS), zero-mean, complex Gaussian random process which possesses the circular property [20, 21], namely $w_m(t)$ and $w_m(t)e^{j\vartheta}$ are identically distributed, $\forall \vartheta \in [0, 2\pi)$, with j denoting, in turn, the imaginary unit.

The Gaussian assumption for the disturbance is justified when clutter arises from the superposition of returns from a large number of equivalent elementary scatterers, independent of one another, whence the resulting process is Gaussian as a consequence of the central limit theorem. However, experimental data indicate that large deviations from the Gaussian model are possible; for instance, deviations are observed for low grazing angles and/or high-resolution radars [22, 23]. It also turns out that conventional detectors, namely the ones designed under the assumption of Gaussian disturbance, suffer inflation of the number of false alarms and/or marked detection performance degradation as the actual amplitude probability density function of the clutter significantly deviates from the Rayleigh law. Nevertheless, non-Gaussian scenarios are out of the scope of this book.

2.2 DISCRETE-TIME TARGET AND NOISE MODELS

Suppose that the radar antenna is steered along the ψ direction and transmits the following coherent burst of RF pulses

$$As(t)\cos(2\pi f_c t + \varphi) = \Re e\left\{Ae^{j\varphi}s(t)e^{j2\pi f_c t}\right\}, \quad t \in [0, N_p T),$$

where $A > 0$ is an amplitude factor related to the transmitted power, $\varphi \in [0, 2\pi)$ is the initial phase of the carrier signal, $\Re e\{z\}$ is the real part of the complex number z, and

$$s(t) = \sum_{n=0}^{N_p-1} p(t - nT),$$

with $p(t)$ a rectangular pulse of duration T_p (T_p is much smaller than T) such that

$$\int_{-\infty}^{+\infty} |p(t)|^2 dt = 1.$$

Assume that the radar illuminates a point-like target moving with constant radial velocity v (with $v > 0$ for a target approaching the radar); then, at time t, the received target echo at the m-th sensor, $m = 1, \ldots, N_a$, is given by [24]

$$\Re e\left\{x_m(t)e^{j2\pi f_c t}\right\} = \Re e\left\{\alpha s(t - \tau(t))e^{j2\pi f_c (t-\tau(t))}e^{j2\pi f_c \Delta_m}\right\}, \tag{2.2}$$

where

- $x_m(t)$ (as already stated) is the complex envelope of the received target echo at the m-th antenna element;
- $\tau(t)$ is the round trip delay (with respect to the origin of the reference system) and $\Delta_m = (m-1)d\cos\psi/c$ is the travel time between the m-th sensor and the origin (where we have neglected the dependence of ψ on t);
- $\alpha \in \mathbb{C}$ is a factor which accounts for $Ae^{j\varphi}$, the effects of the transmitting antenna gain, the radiation pattern of the array sensors, the two-way path loss, and the radar cross-section of the (slowly-fluctuating) target.

In order to determine the useful target component, observe that a signal received at time t has been reflected at $t - \tau(t)/2$ and that the corresponding target range (i.e., the distance from the origin) is

$$R\left(t - \frac{\tau(t)}{2}\right) = R(0) - v\left(t - \frac{\tau(t)}{2}\right) = \frac{c\tau(t)}{2},$$

where $R(t)$ is the range at time t (and $R(0)$ is the initial range). Solving the above equation with respect to $\tau(t)$ (ignoring relativistic effects) yields

$$\tau(t) = \frac{2R(0)}{c-v} - \frac{2vt}{c-v} \approx \frac{2R(0)}{c} - \frac{2vt}{c} = \tau - \frac{2v}{c}t, \tag{2.3}$$

where $\tau = 2R(0)/c$. By inserting (2.3) into (2.2), we obtain

$$\Re e\left\{x_m(t)e^{j2\pi f_c t}\right\} = \Re e\left\{\alpha s\left(t\left(1+\frac{2v}{c}\right)-\tau\right)e^{j2\pi(f_c+f_d)t}e^{j2\pi(m-1)\nu_s}e^{-j2\pi f_c\tau}\right\}, \tag{2.4}$$

where

$$f_d = f_c\frac{2v}{c}$$

is the Doppler frequency shift of the target backscattered signal and ν_s is the target *spatial frequency*, given by

$$\nu_s = \frac{d}{\lambda}\cos\psi. \tag{2.5}$$

Inspection of equation (2.4) highlights that the target velocity compresses or stretches the time scale of the transmitted pulses. However, the above effect is negligible for realistic target velocities and usual radar parameters. As a consequence, after complex baseband conversion, the target signal at the m-th antenna element is given by

$$\begin{aligned} x_m(t) &= \alpha e^{-j2\pi f_c\tau} s(t-\tau)\, e^{j2\pi f_d t} e^{j2\pi(m-1)\nu_s} \\ &= \alpha \sum_{n=0}^{N_p-1} p(t-nT-\tau) e^{j2\pi f_d t} e^{j2\pi(m-1)\nu_s}, \end{aligned}$$

where α in the right-most side of previous equation incorporates $e^{-j2\pi f_c\tau}$. Hereafter α will denote any proper modification of the original constant.

A discrete form for the received signal at the m-th sensor is obtained by sampling the output of a filter matched[4] to $p(t)$ and fed by $r_m(t)$ as illustrated in Figure 2.2. Due to the superposition

[4]The Doppler shift makes the filter mismatched to the incoming signal, but we can reasonably assume that the expected range of Doppler frequencies is much smaller than the one-sided pulse bandwidth W_p, i.e.,

$$W_p \approx \frac{1}{T_p} \gg f_d = f_c\frac{2v}{c},$$

and neglect the frequency mismatch. For instance, assuming $v = 1032$ m/sec (Mach-3), $T_p = 0.2$ μsec, and $f_c = 10^9$ Hz yields

$$W_p \approx 5 \text{ MHz} \gg f_d = 6.88 \text{ kHz}.$$

principle, we can leave aside for the moment the noise component and focus on the discrete form of the useful component.

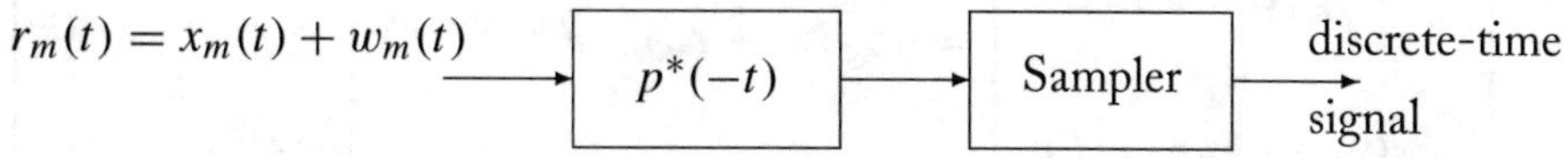

Figure 2.2: Discrete-time representation of the complex envelope of the received signal at the m-th antenna element.

In particular, the matched filter output for the m-th sensor is given by

$$\begin{aligned} y_m(t) &= x_m(t) * p^*(-t) = \int_{-\infty}^{+\infty} x_m(u) p^*(u-t) du \\ &= \int_{-\infty}^{+\infty} \alpha s\,(u-\tau)\, e^{j2\pi f_d u} e^{j2\pi(m-1)\nu_s} p^*(u-t) du \\ &= \alpha e^{j2\pi(m-1)\nu_s} \int_{-\infty}^{+\infty} \sum_{n=0}^{N_p-1} p(u-nT-\tau) p^*(u-t) e^{j2\pi f_d u} du. \end{aligned}$$

Let $u_1 = u - nT - \tau$, then

$$\begin{aligned} y_m(t) &= \alpha e^{j2\pi(m-1)\nu_s} \sum_{n=0}^{N_p-1} \int_{-\infty}^{+\infty} p(u_1) p^*(u_1 - (t-nT-\tau)) e^{j2\pi f_d (u_1+nT+\tau)} du_1 \\ &= \alpha e^{j2\pi(m-1)\nu_s} \sum_{n=0}^{N_p-1} e^{j2\pi f_d(\tau+nT)} \underbrace{\int_{-\infty}^{+\infty} p(u_1) p^*(u_1-(t-nT-\tau)) e^{j2\pi f_d u_1} du_1}_{\chi_p(t-nT-\tau, f_d)} \\ &= \alpha e^{j2\pi f_d \tau} e^{j2\pi(m-1)\nu_s} \sum_{n=0}^{N_p-1} \chi_p(t-nT-\tau,\, f_d) e^{j2\pi n f_d T} \\ &= \alpha e^{j2\pi(m-1)\nu_s} \sum_{n=0}^{N_p-1} \chi_p(t-nT-\tau,\, f_d) e^{j2\pi n \nu_d}, \end{aligned}$$

where $\chi_p(\cdot,\cdot)$ is the *ambiguity function* of the pulse waveform $p(t)$ [25] and

$$\nu_d = f_d T \tag{2.6}$$

is the *normalized target Doppler frequency*.

In order to generate the range gate corresponding to a round-trip delay τ_c, $y_m(t)$ is sampled at the time instants $t_i = \tau_c + (i-1)T$, $i = 1, \ldots, N_p$; it is easy to verify that, for a point-like target located in this range gate, the N_p-dimensional vector whose i-th entry is given by $y_m(t_i)$, $i = 1, \ldots, N_p$, can be recast as[5]

$$\boldsymbol{y}_m = \begin{bmatrix} y_m(\tau_c) \\ y_m(\tau_c + T) \\ \vdots \\ y_m(\tau_c + (N_p - 1)T) \end{bmatrix} = \alpha \chi_p(\tau_c - \tau, f_d) e^{j2\pi(m-1)\nu_s} \begin{bmatrix} 1 \\ e^{j2\pi\nu_d} \\ \vdots \\ e^{j2\pi(N_p-1)\nu_d} \end{bmatrix}$$

$$= \alpha e^{j2\pi(m-1)\nu_s} \boldsymbol{s}(\nu_d) \in \mathbb{C}^{N_p \times 1}, \quad m = 1, \ldots, N_a, \tag{2.7}$$

where

$$\boldsymbol{s}(\nu_d) = \begin{bmatrix} 1 \\ e^{j2\pi\nu_d} \\ \vdots \\ e^{j2\pi(N_p-1)\nu_d} \end{bmatrix} \tag{2.8}$$

is the *temporal steering vector*. Now let us consider the overall data matrix

$$\boldsymbol{Y} = \begin{bmatrix} \boldsymbol{y}_1^T \\ \boldsymbol{y}_2^T \\ \vdots \\ \boldsymbol{y}_{N_a}^T \end{bmatrix} = \alpha \begin{bmatrix} \boldsymbol{s}(\nu_d)^T \\ e^{j2\pi\nu_s} \boldsymbol{s}(\nu_d)^T \\ \vdots \\ e^{j2\pi(N_a-1)\nu_s} \boldsymbol{s}(\nu_d)^T \end{bmatrix} \in \mathbb{C}^{N_a \times N_p},$$

where ^{T} denotes transpose, and generate the N-dimensional vector $\boldsymbol{y}$, with $N = N_a N_p$, by stacking the columns of $\boldsymbol{Y}$, i.e.,

$$\boldsymbol{y} = \text{vec}(\boldsymbol{Y}) = \alpha \boldsymbol{s}(\nu_d) \otimes \boldsymbol{a}(\nu_s) = \alpha \boldsymbol{v}(\nu_d, \nu_s) \in \mathbb{C}^{N \times 1}, \tag{2.9}$$

where $\otimes$ is the Kronecker product,

$$\boldsymbol{a}(\nu_s) = \begin{bmatrix} 1 \\ e^{j2\pi\nu_s} \\ \vdots \\ e^{j2\pi(N_a-1)\nu_s} \end{bmatrix} \in \mathbb{C}^{N_a \times 1} \tag{2.10}$$

is usually named *spatial steering vector*, and $\boldsymbol{v}(\nu_d, \nu_s)$ is the overall *space-time steering vector*. Note that the general expression for the steering vector encompasses the following special cases

$$\begin{cases} \boldsymbol{v}(\nu_d, \nu_s) \equiv \boldsymbol{s}(\nu_d), & \text{if } N_a = 1, \\ \boldsymbol{v}(\nu_d, \nu_s) \equiv \boldsymbol{a}(\nu_s), & \text{if } N_p = 1. \end{cases}$$

Table 2.1 provides a list of the radar parameters involved in the considered framework.

[5]It is also necessary to assume that adjacent range gates do not contain targets and remember that $\chi_p(\tau_c + (i - n)T - \tau, f_d) = 0$, $\forall n \neq i$.

Table 2.1: Radar parameters.

N_a	number of sensors
N_p	number of pulses
T_p	pulse duration
T	PRT
f_c	radar carrier frequency
λ	radar operating wavelength
d	interelement spacing
W_p	one-sided pulse bandwidth
ν_d	normalized target Doppler frequency
ν_s	target spatial frequency

As to the noise component at the output of the m-th sampler, it is given by

$$\boldsymbol{n}_m = \begin{bmatrix} n_m(\tau_c) \\ n_m(\tau_c + T) \\ \vdots \\ n_m(\tau_c + (N_p - 1)T) \end{bmatrix} \in \mathbb{C}^{N_p \times 1}, \quad m = 1, \ldots, N_a,$$

where

$$n_m(t) = \int_{-\infty}^{+\infty} w_m(u) p^*(u - t) du$$

is also a circularly symmetric WSS complex Gaussian random process. The overall noise matrix is

$$\boldsymbol{N} = \begin{bmatrix} \boldsymbol{n}_1^T \\ \boldsymbol{n}_2^T \\ \vdots \\ \boldsymbol{n}_{N_a}^T \end{bmatrix} \in \mathbb{C}^{N_a \times N_p}.$$

Again, by stacking the columns of $\boldsymbol{N}$, we obtain

$$\boldsymbol{n} = \mathrm{vec}(\boldsymbol{N}) \in \mathbb{C}^{N \times 1}.$$

In the sequel we assume that $\boldsymbol{n}$ is a multivariate complex normal vector with zero mean and (positive definite) covariance matrix $\boldsymbol{M} \in \mathbb{C}^{N \times N}$ and write $\boldsymbol{n} \sim \mathcal{CN}_N(\boldsymbol{0}, \boldsymbol{M})$ (for further details see Appendix A).

According to the superposition principle, the discrete-time counterpart of equation (2.1) is

$$\boldsymbol{r} = \alpha \boldsymbol{v}(\nu_d, \nu_s) + \boldsymbol{n}.$$

The next section is aimed at reviewing those tools of detection theory explicitly used in this book.

2.3 THE GLRT AND THE TWO-STEP GLRT-BASED DESIGN PROCEDURE WITH APPLICATION TO ADAPTIVE DETECTION OF A SIGNAL KNOWN UP TO A COMPLEX FACTOR

Radar detection theory is primarily aimed at solving binary hypothesis tests by resorting to reliable "decision rules" (or detectors) that, based on the received signal $\boldsymbol{r}$, decide whether it contains noise only, i.e., $\boldsymbol{r} = \boldsymbol{n}$ (H_0 hypothesis) or signal plus noise, i.e., $\boldsymbol{r} = \alpha \boldsymbol{v}(\nu_d, \nu_s) + \boldsymbol{n}$ (H_1 hypothesis). There exist several excellent books addressing detection theory and its applications as, for instance, [26, 27, 28]. Herein, we only focus on a specific class of detection problems tied to the target and noise models of previous section and, as a preliminary step, we observe that such models contain several nuisance parameters that require further attention.

As to the noise, the covariance matrix $\boldsymbol{M}$ is in general not known and, hence, it must be estimated from the data. To this end, several approaches are possible; the most commonly used are the following:

- modeling the noise as the output of an autoregressive filter [29, 30, 31];
- assuming that a set of *secondary data* $\boldsymbol{r}_k = \boldsymbol{n}_k, k = 1, \ldots, K$, namely a set of returns which are supposed free of signal components, but sharing the same spectral properties of the noise in the cell under test[6] (CUT), is available [2]. This scenario is usually referred to as *homogeneous environment*. Alternatively, it is also reasonable to consider the so-called *partially homogeneous environment* [16] where the covariance matrix of the CUT and that of secondary data coincide only up to a scaling factor. More general scenarios could also be considered.

In the following we assume that a set of $K \geq N$ secondary data is available: such data could be obtained by processing range gates in spatial proximity with that under test. Moreover, we consider the homogeneous and the partially homogeneous environments. More general scenarios would make the design of adaptive detection schemes a formidable problem and are out of the scope of this book. The interested reader is referred to [32, and references therein] for possible viable means to circumvent the "heterogeneity" of the environment.

As to the signal, in the remaining part of this chapter we test the presence of a target within a given space/time radar cell; to this end, we assume that the parameters ν_d and ν_s and, eventually, the target steering vector, are perfectly known[7]: ways to relax this assumption are the object of this book and will be developed in subsequent chapters.

Several models can be used for the complex factor α. More specifically, the following instances come in handy:

[6] Also referred to in the following as primary data.
[7] For this reason such parameters are omitted.

1. α is an unknown deterministic parameter;

2. the amplitude is an unknown deterministic parameter while the phase is a random variable (rv) uniformly distributed in $(0, 2\pi)$;

3. the amplitude is a rv ruled by a known probability density function (pdf) or an unknown pdf belonging to a known family of pdf's, while the phase is a rv uniformly distributed in $(0, 2\pi)$ and independent of the former.

In the following we assume that α is an unknown deterministic parameter.

Summarizing, the detection problem at hand is

$$\begin{cases} H_0 : \begin{cases} \boldsymbol{r} = \boldsymbol{n}, \\ \boldsymbol{r}_k = \boldsymbol{n}_k, \qquad k = 1, \ldots, K, \end{cases} \\ \\ H_1 : \begin{cases} \boldsymbol{r} = \alpha \boldsymbol{v} + \boldsymbol{n}, \\ \boldsymbol{r}_k = \boldsymbol{n}_k, \qquad k = 1, \ldots, K, \end{cases} \end{cases} \tag{2.11}$$

where

- $\boldsymbol{n}$, $\boldsymbol{n}_k$, $k = 1, \ldots, K \geq N$, are independent, zero-mean, complex normal random vectors; moreover, we assume that they share either one and the same (unknown) covariance matrix $\boldsymbol{M}$, i.e., $\boldsymbol{n}, \boldsymbol{n}_k \sim \mathcal{CN}_N(\boldsymbol{0}, \boldsymbol{M})$ (homogeneous environment) or the same covariance matrix up to a scale factor, i.e., $\boldsymbol{n} \sim \mathcal{CN}_N(\boldsymbol{0}, \sigma^2 \boldsymbol{M})$, $\boldsymbol{n}_k \sim \mathcal{CN}_N(\boldsymbol{0}, \boldsymbol{M})$ (partially homogeneous environment[8]);

- $\alpha \in \mathbb{C}$ is an unknown deterministic parameter and $\boldsymbol{v}$ is for the moment a perfectly known vector.

The design of reliable decision rules on the hypothesis actually in force requires introducing meaningful performance parameters. To this end, note that a binary decision rule may incur into two type of errors, namely

- a false alarm error corresponding to the case that the hypothesis H_0 is in force, but the decision rule selects H_1;

- a miss error corresponding to the case that the hypothesis H_1 is in force, but the decision rule selects H_0.

Accordingly, for a given decision rule, namely a mapping from the space of the observables $\boldsymbol{r}$, $\boldsymbol{r}_k$, $k = 1, \ldots K$, to the set $\{H_0, H_1\}$, we define

[8]The partially homogeneous environment is used only to derive the ACE; throughout the book the analysis will assume the homogeneous (Gaussian) environment.

- the false alarm probability (P_{fa}) as the supremum of the probability of accepting H_1 under H_0 (over the unknown parameters of the distribution of the observables under H_0); if the decision rule has a distribution under H_0 independent of the nuisance parameters we say that the decision rule guarantees the CFAR property and the evaluation of the supremum is no longer necessary;
- the probability of detection (P_d) as the probability of accepting H_1 when it is actually in force; note that P_d is one minus the miss error probability and that it depends on the (possibly unknown) distributional parameters under H_1.

For the case at hand a uniformly most powerful test, namely a decision rule that maximizes P_d (regardless of the unknown parameters of the distribution of the observables under H_1), for a preassigned P_{fa}, does not exist. For this reason we resort to the GLRT; it relies on the generalized likelihood ratio (GLR) statistic, namely the ratio between the likelihood (pdf) of the observables under H_1 and that under H_0 with the unknown parameters of the pdf's substituted by their maximum likelihood (ML) estimates [26]. The GLRT compares the GLR statistic, $t(\boldsymbol{r}, \boldsymbol{r}_1, \ldots, \boldsymbol{r}_K)$ say, with a proper threshold, η say, selecting H_1 when the statistic is above the threshold, H_0 otherwise; in symbols we write

$$t(\boldsymbol{r}, \boldsymbol{r}_1, \ldots, \boldsymbol{r}_K) \underset{H_0}{\overset{H_1}{\gtrless}} \eta.$$

More specifically, the GLRT for homogeneous environment is given by

$$\frac{\max\limits_{\alpha} \max\limits_{\boldsymbol{M}} f(\boldsymbol{r}, \boldsymbol{r}_1, \ldots, \boldsymbol{r}_K; \alpha, \boldsymbol{M})}{\max\limits_{\boldsymbol{M}} f(\boldsymbol{r}, \boldsymbol{r}_1, \ldots, \boldsymbol{r}_K; \boldsymbol{M})} \underset{H_0}{\overset{H_1}{\gtrless}} \eta, \tag{2.12}$$

where $f(\boldsymbol{r}, \boldsymbol{r}_1, \ldots, \boldsymbol{r}_K; \alpha, \boldsymbol{M})$ and $f(\boldsymbol{r}, \boldsymbol{r}_1, \ldots, \boldsymbol{r}_K; \boldsymbol{M})$ denote the likelihood functions under the H_1 and the H_0 hypothesis, respectively, and η is the threshold to be set in order to guarantee a preassigned P_{fa}. Analogously, the GLRT for partially homogeneous environment can be expressed as

$$\frac{\max\limits_{\alpha} \max\limits_{\sigma^2, \boldsymbol{M}} f(\boldsymbol{r}, \boldsymbol{r}_1, \ldots, \boldsymbol{r}_K; \alpha, \sigma^2, \boldsymbol{M})}{\max\limits_{\sigma^2, \boldsymbol{M}} f(\boldsymbol{r}, \boldsymbol{r}_1, \ldots, \boldsymbol{r}_K; \sigma^2, \boldsymbol{M})} \underset{H_0}{\overset{H_1}{\gtrless}} \eta, \tag{2.13}$$

with $f(\boldsymbol{r}, \boldsymbol{r}_1, \ldots, \boldsymbol{r}_K; \alpha, \sigma^2, \boldsymbol{M})$ and $f(\boldsymbol{r}, \boldsymbol{r}_1, \ldots, \boldsymbol{r}_K; \sigma^2, \boldsymbol{M})$ denoting the likelihood functions for this scenario.

Another way to attack problem (2.11) is to resort to an ad hoc procedure based on the so-called two-step GLRT-based design procedure [6]. To be more precise, denote by $f(\boldsymbol{r}; \alpha, \boldsymbol{M})$

($f(\boldsymbol{r}; \alpha, \sigma^2, \boldsymbol{M})$) and $f(\boldsymbol{r}; \boldsymbol{M})$ ($f(\boldsymbol{r}; \sigma^2, \boldsymbol{M})$) the pdf's of $\boldsymbol{r}$ under H_1 and H_0, respectively, in homogeneous (partially homogeneous) environment. The following rationale can be adopted

- first, assume that $\boldsymbol{M}$ is known and implement the GLRT, given by

$$\frac{\max_{\alpha} f(\boldsymbol{r}; \alpha, \boldsymbol{M})}{f(\boldsymbol{r}; \boldsymbol{M})} \underset{H_0}{\overset{H_1}{\gtrless}} \eta \tag{2.14}$$

  and

$$\frac{\max_{\alpha} \max_{\sigma^2} f(\boldsymbol{r}; \alpha, \sigma^2, \boldsymbol{M})}{\max_{\sigma^2} f(\boldsymbol{r}; \sigma^2, \boldsymbol{M})} \underset{H_0}{\overset{H_1}{\gtrless}} \eta, \tag{2.15}$$

  for homogeneous and partially homogeneous environment, respectively;

- then, replace the unknown matrix $\boldsymbol{M}$ with a proper estimate; a reasonable choice is the sample covariance matrix based on the secondary data set, i.e.,

$$\widehat{\boldsymbol{M}} = \frac{1}{K} \sum_{k=1}^{K} \boldsymbol{r}_k \boldsymbol{r}_k^\dagger,$$

  where † denotes conjugate transpose.

GLRT and ad hoc detectors for the hypothesis testing problem (2.11) will be the object of the next section.

2.3.1 ADAPTIVE DETECTION OF A SIGNAL KNOWN UP TO A COMPLEX FACTOR

For the reader's ease we briefly review the derivation of the GLRT for the hypothesis testing problem (2.11) and homogeneous environment. We will follow the lead of the original derivation proposed by Kelly in [2].

As a first step towards the derivation of the GLRT we have to specify the pdf's of the observables under both hypotheses. We have that

$$\begin{aligned} f(\boldsymbol{r}, \boldsymbol{r}_1, \ldots, \boldsymbol{r}_K; \alpha, \boldsymbol{M}) &= \frac{1}{\pi^{N(K+1)}} \frac{1}{\det^{K+1}(\boldsymbol{M})} \\ &\times e^{-\operatorname{tr}\left\{\boldsymbol{M}^{-1}\left[(\boldsymbol{r} - \alpha \boldsymbol{v})(\boldsymbol{r} - \alpha \boldsymbol{v})^\dagger + \sum_{k=1}^{K} \boldsymbol{r}_k \boldsymbol{r}_k^\dagger\right]\right\}} \end{aligned}$$

under H_1 and

$$f(\boldsymbol{r}, \boldsymbol{r}_1, \ldots, \boldsymbol{r}_K; \boldsymbol{M}) = \frac{1}{\pi^{N(K+1)}} \frac{1}{\det^{K+1}(\boldsymbol{M})} e^{-\operatorname{tr}\left\{\boldsymbol{M}^{-1}\left[\boldsymbol{r}\boldsymbol{r}^\dagger + \sum_{k=1}^{K} \boldsymbol{r}_k \boldsymbol{r}_k^\dagger\right]\right\}}$$

under H_0, respectively, with $\det(\cdot)$ and $\mathrm{tr}\{\cdot\}$ denoting the determinant and the trace of the matrix argument.

Moreover, observe that

$$\arg\max_{\boldsymbol{M}} f(\boldsymbol{r}, \boldsymbol{r}_1, \ldots, \boldsymbol{r}_K; \alpha, \boldsymbol{M}) = \frac{1}{K+1}\left[(\boldsymbol{r} - \alpha\boldsymbol{v})(\boldsymbol{r} - \alpha\boldsymbol{v})^\dagger + \sum_{k=1}^{K} \boldsymbol{r}_k \boldsymbol{r}_k^\dagger\right]$$

and

$$\arg\max_{\boldsymbol{M}} f(\boldsymbol{r}, \boldsymbol{r}_1, \ldots, \boldsymbol{r}_K; \boldsymbol{M}) = \frac{1}{K+1}\left[\boldsymbol{r}\boldsymbol{r}^\dagger + \sum_{k=1}^{K} \boldsymbol{r}_k \boldsymbol{r}_k^\dagger\right].$$

Inserting the above expressions for $\boldsymbol{M}$ into equation (2.12) yields

$$\max_{\alpha}\left[\frac{\det\left(\boldsymbol{r}\boldsymbol{r}^\dagger + \sum_{k=1}^{K} \boldsymbol{r}_k \boldsymbol{r}_k^\dagger\right)}{\det\left((\boldsymbol{r} - \alpha\boldsymbol{v})(\boldsymbol{r} - \alpha\boldsymbol{v})^\dagger + \sum_{k=1}^{K} \boldsymbol{r}_k \boldsymbol{r}_k^\dagger\right)}\right]^{K+1} \underset{H_0}{\overset{H_1}{\gtrless}} \eta$$

that can be equivalently rewritten as

$$\frac{\det\left(\boldsymbol{r}\boldsymbol{r}^\dagger + \sum_{k=1}^{K} \boldsymbol{r}_k \boldsymbol{r}_k^\dagger\right)}{\min_{\alpha}\det\left((\boldsymbol{r} - \alpha\boldsymbol{v})(\boldsymbol{r} - \alpha\boldsymbol{v})^\dagger + \sum_{k=1}^{K} \boldsymbol{r}_k \boldsymbol{r}_k^\dagger\right)} \underset{H_0}{\overset{H_1}{\gtrless}} \eta, \qquad (2.16)$$

where, hereafter, η denotes any modification of the original threshold.

In order to accomplish the required maximization with respect to α let

$$\boldsymbol{S} = \sum_{k=1}^{K} \boldsymbol{r}_k \boldsymbol{r}_k^\dagger$$

and denote by[9] $\boldsymbol{S}^{-1/2}$ the (positive definite Hermitian) square root of $\boldsymbol{S}^{-1}$. It follows that

$$\begin{aligned}
\det\left((\boldsymbol{r} - \alpha\boldsymbol{v})(\boldsymbol{r} - \alpha\boldsymbol{v})^\dagger + \boldsymbol{S}\right) &= \det(\boldsymbol{S})\det\left(\boldsymbol{S}^{-1/2}(\boldsymbol{r} - \alpha\boldsymbol{v})(\boldsymbol{r} - \alpha\boldsymbol{v})^\dagger\boldsymbol{S}^{-1/2} + \boldsymbol{I}\right) \\
&= \det(\boldsymbol{S})\det\left((\boldsymbol{r}_S - \alpha\boldsymbol{v}_S)(\boldsymbol{r}_S - \alpha\boldsymbol{v}_S)^\dagger + \boldsymbol{I}\right) \\
&= \det(\boldsymbol{S})\left[1 + (\boldsymbol{r}_S - \alpha\boldsymbol{v}_S)^\dagger(\boldsymbol{r}_S - \alpha\boldsymbol{v}_S)\right] \\
&= \det(\boldsymbol{S})\left[1 + \|\boldsymbol{r}_S - \alpha\boldsymbol{v}_S\|^2\right],
\end{aligned}$$

[9]Note that the positive semidefinite Hermitian matrix $\boldsymbol{S}$ is nonsingular and, hence, invertible, with probability one if $K \geq N$ [33] (see also Appendix A).

where $\boldsymbol{r}_S = \boldsymbol{S}^{-1/2}\boldsymbol{r}$, $\boldsymbol{v}_S = \boldsymbol{S}^{-1/2}\boldsymbol{v}$, $\boldsymbol{I}$ is the identity matrix of proper dimensions, $\|\cdot\|$ is the Euclidean norm of a vector, and the second-last equality follows from identity

$$\det(\boldsymbol{I} + \boldsymbol{B}\boldsymbol{C}) = \det(\boldsymbol{I} + \boldsymbol{C}\boldsymbol{B}), \tag{2.17}$$

with $\boldsymbol{B} \in \mathbb{C}^{r\times s}$ and $\boldsymbol{C} \in \mathbb{C}^{s\times r}$, $r, s \in \mathbb{N}$, arbitrary matrices [34].

Thus, we have that

$$\begin{aligned}\min_{\alpha} \det\left((\boldsymbol{r} - \alpha\boldsymbol{v})(\boldsymbol{r} - \alpha\boldsymbol{v})^\dagger + \boldsymbol{S}\right) &= \min_{\alpha} \det(\boldsymbol{S})\left[1 + \|\boldsymbol{r}_S - \alpha\boldsymbol{v}_S\|^2\right] \\ &= \det(\boldsymbol{S})\left[1 + \|\boldsymbol{P}^{\perp}_{\boldsymbol{v}_S}\boldsymbol{r}_S\|^2\right], \end{aligned} \tag{2.18}$$

where $\boldsymbol{P}^{\perp}_{\boldsymbol{v}_S}$ denotes the projection matrix onto the orthogonal complement of the space spanned by $\boldsymbol{v}_S$, namely

$$\boldsymbol{P}^{\perp}_{\boldsymbol{v}_S} = \boldsymbol{I} - \boldsymbol{v}_S(\boldsymbol{v}_S^\dagger\boldsymbol{v}_S)^{-1}\boldsymbol{v}_S^\dagger = \boldsymbol{I} - \boldsymbol{S}^{-1/2}\boldsymbol{v}(\boldsymbol{v}^\dagger\boldsymbol{S}^{-1}\boldsymbol{v})^{-1}\boldsymbol{v}^\dagger\boldsymbol{S}^{-1/2}.$$

Similarly, it is easy to show that

$$\det\left(\boldsymbol{r}\boldsymbol{r}^\dagger + \sum_{k=1}^{K}\boldsymbol{r}_k\boldsymbol{r}_k^\dagger\right) = \det(\boldsymbol{S})\left(1 + \boldsymbol{r}_S^\dagger\boldsymbol{r}_S\right). \tag{2.19}$$

Replacing the numerator and the denominator of equation (2.16) with the right-hand side of equations (2.19) and (2.18), respectively, yields

$$\frac{1 + \boldsymbol{r}_S^\dagger\boldsymbol{r}_S}{1 + \boldsymbol{r}_S^\dagger\boldsymbol{P}^{\perp}_{\boldsymbol{v}_S}\boldsymbol{r}_S} = \frac{1 + \boldsymbol{r}^\dagger\boldsymbol{S}^{-1}\boldsymbol{r}}{1 + \boldsymbol{r}^\dagger\boldsymbol{S}^{-1}\boldsymbol{r} - \dfrac{|\boldsymbol{r}^\dagger\boldsymbol{S}^{-1}\boldsymbol{v}|^2}{\boldsymbol{v}^\dagger\boldsymbol{S}^{-1}\boldsymbol{v}}} \underset{H_0}{\overset{H_1}{\gtrless}} \eta, \tag{2.20}$$

where we have used the fact that $\boldsymbol{P}^{\perp}_{\boldsymbol{v}_S}$ is Hermitian and idempotent.

Finally, it is straightforward to recast the test in the well-known form [2]

$$t_{\mathrm{K}} = \frac{|\boldsymbol{r}^\dagger\boldsymbol{S}^{-1}\boldsymbol{v}|^2}{(\boldsymbol{v}^\dagger\boldsymbol{S}^{-1}\boldsymbol{v})(1 + \boldsymbol{r}^\dagger\boldsymbol{S}^{-1}\boldsymbol{r})} \underset{H_0}{\overset{H_1}{\gtrless}} \eta. \tag{2.21}$$

It is also worth deriving the GLRT for the hypothesis test (2.11), but assuming a partially homogeneous environment. For this case the pdf's of the observables are given by

$$\begin{aligned} f(\boldsymbol{r}, \boldsymbol{r}_1, \ldots, \boldsymbol{r}_K; \alpha, \sigma^2, \boldsymbol{M}) &= \frac{1}{\pi^{N(K+1)}}\frac{1}{\sigma^{2N}\det^{K+1}(\boldsymbol{M})} \\ &\times e^{-\operatorname{tr}\left\{\boldsymbol{M}^{-1}\left[\frac{1}{\sigma^2}(\boldsymbol{r} - \alpha\boldsymbol{v})(\boldsymbol{r} - \alpha\boldsymbol{v})^\dagger + \sum_{k=1}^{K}\boldsymbol{r}_k\boldsymbol{r}_k^\dagger\right]\right\}} \end{aligned}$$

under H_1 and

$$f(\boldsymbol{r}, \boldsymbol{r}_1, \ldots, \boldsymbol{r}_K; \sigma^2, \boldsymbol{M}) = \frac{1}{\pi^{N(K+1)}} \frac{1}{\sigma^{2N} \det^{K+1}(\boldsymbol{M})} e^{-\mathrm{tr}\left\{\boldsymbol{M}^{-1}\left[\frac{1}{\sigma^2}\boldsymbol{r}\boldsymbol{r}^\dagger + \sum_{k=1}^{K} \boldsymbol{r}_k \boldsymbol{r}_k^\dagger\right]\right\}}$$

under H_0.

Following the lead of [2] it is not difficult to show that the GLRT for partially homogeneous environment, given by equation (2.13), yields the ACE [16], also known as adaptive normalized matched filter (ANMF) [15], i.e.,

$$t_{\mathrm{ACE}} = \frac{|\boldsymbol{r}^\dagger \boldsymbol{S}^{-1} \boldsymbol{v}|^2}{(\boldsymbol{v}^\dagger \boldsymbol{S}^{-1} \boldsymbol{v})(\boldsymbol{r}^\dagger \boldsymbol{S}^{-1} \boldsymbol{r})} \underset{H_0}{\overset{H_1}{\gtrless}} \eta. \tag{2.22}$$

The derivation of the ad hoc detectors for the hypothesis test (2.11) and either homogeneous or partially homogeneous environment is straightforward and is left as exercise to the interested reader. It turns out that the ad hoc detectors are the AMF [6], namely

$$t_{\mathrm{AMF}} = \frac{|\boldsymbol{r}^\dagger \boldsymbol{S}^{-1} \boldsymbol{v}|^2}{\boldsymbol{v}^\dagger \boldsymbol{S}^{-1} \boldsymbol{v}} \underset{H_0}{\overset{H_1}{\gtrless}} \eta, \tag{2.23}$$

and the ACE (2.22) for homogeneous and partially homogeneous environment, respectively.

As a final comment, observe that if, under the H_0 hypothesis, the distribution of a decision statistic t is independent of the unknown parameters, the detector possesses the CFAR property. The importance of the CFAR property from a practical viewpoint should be apparent: it allows to set the threshold in order to operate at the desired false alarm rate and not at some value of the probability of deciding H_1 when H_0 is in force, lower than the preassigned P_{fa} (equating the preassigned one in the "worst-case" only). It is possible to show that Kelly's detector and AMF guarantee the CFAR property with respect to $\boldsymbol{M}$ in homogeneous environment and that the ACE guarantees the CFAR property with respect to $\boldsymbol{M}$ in homogeneous environment and with respect to σ^2 and $\boldsymbol{M}$ in the partially homogeneous one. The interested reader is referred to [2, 6, 8, 17] and to Appendix B for the performance assessment of the above detectors.

An example of the behavior of Kelly's detector, AMF, and ACE is given in Figure 2.3 where we plot P_d vs the signal-to-noise ratio (SNR), defined as

$$\mathrm{SNR} = |\alpha^2| \boldsymbol{v}^\dagger \boldsymbol{M}^{-1} \boldsymbol{v}, \tag{2.24}$$

assuming $P_{fa} = 10^{-6}$, $N = 16$, homogeneous environment, $K = 32, 48$. For comparison purposes the GLRT for known $\boldsymbol{M}$ is plotted too. Curves reported in Figure 2.3 have been obtained by means

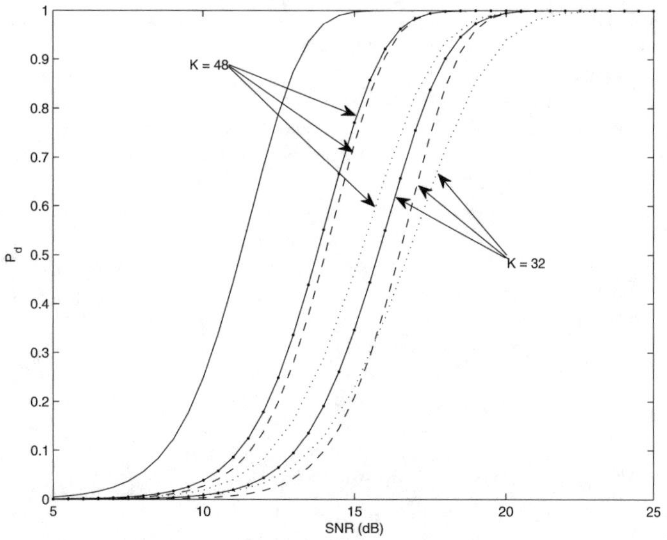

Figure 2.3: P_d vs SNR for the GLRT for known $\boldsymbol{M}$ (solid line), Kelly's detector (solid lines with dot marker), AMF (dashed lines), and ACE (dotted lines), $P_{fa} = 10^{-6}$, $N = 16$, K as parameter.

of numerical integration techniques (see results contained in Appendix B for further details on the statistical characterization of the detectors considered in this chapter).

The horizontal displacement of Kelly's detector with respect to the GLRT for known $\boldsymbol{M}$ is the penalty for having to estimate $\boldsymbol{M}$. The comparison shows that Kelly's detector generally outperforms its competitors (and, remarkably, it will also prove rather selective). For this reason it will be used as a benchmark for performance comparisons assuming matched signals.

A few final remarks are in order. First, observe that the non-negligible loss of the ACE with respect to Kelly's detector is not surprising since the former has been derived assuming partially homogeneous environment[10]. It will be shown in Chapter 3 that resorting to the ACE in homogeneous environment guarantees an increased capability to reject mismatched signals, namely coherent echoes from a direction different to that in which the radar is steered. Second, the hypothesis testing problem (2.11) can be generalized to the case of signals belonging to a preassigned subspace of the observables [8], in order to increase robustness to slightly mismatched signals, see also Chapter 3. Finally, it is worth stressing that the above adaptive detection schemes, as well as their generalizations to detect subspace signals, can be used in several other contexts as, for example, to detect acoustic signals in active sonars [35] and multidimensional signals in hyperspectral imaging [36].

[10] Actually, the ACE has been firstly derived as an approximation to the GLRT to detect signals known up to a multiplicative factor in non-Gaussian noise modeled as a compound-Gaussian one [15].

CHAPTER 3

Adaptive Detection Schemes for Mismatched Signals

In realistic scenarios the direction of the signal backscattered from a target illuminated by the main beam of the radar antenna may be different from the nominal steering vector due to several environmental and/or instrumental factors. Moreover, sidelobe signals can arise due to the presence of coherent repeater interference or to strong targets that could trigger a detection even when located outside the main beam. Conventional adaptive detection algorithms, namely those designed under the assumption of perfect match between the nominal and the actual steering vector, behave quite differently in presence of mismatched signals. In this chapter we show how to increase the robustness or the selectivity of a decision scheme and, eventually, how to design tunable receivers. It is worth pointing out that all of the detectors that will be considered throughout the chapter guarantee the CFAR property with respect to the noise covariance matrix in homogeneous environment.

3.1 MISMATCHED SIGNALS

In practice, equation (2.9) may not correctly model the actual structure of the backscattered signal. In fact, the presence of mismatched signals arises due to several reasons as, for example [4, 5]

- coherent scattering received from a direction different to that in which the radar system is steered, namely from a sidelobe interferer (see Figure 3.1);
- imperfect modeling of the mainlobe target by the nominal steering vector, where the mismatch may be due to multipath propagation effects, array calibration uncertainties, and beampointing errors (see Figure 3.1).

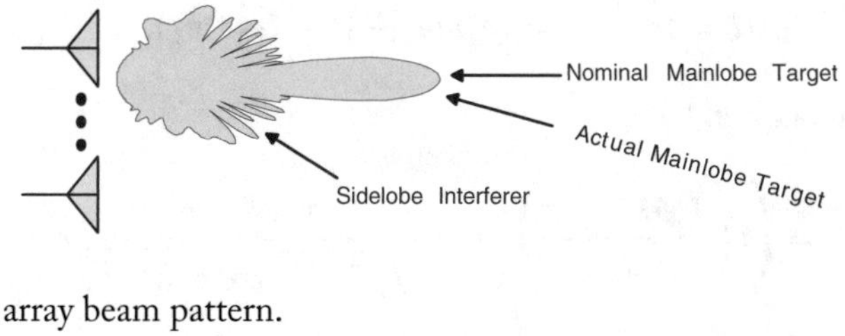

Figure 3.1: Antenna array beam pattern.

Hence, from a general point of view, coherent returns may lie along the direction of a vector $\boldsymbol{p}$, not necessarily aligned with $\boldsymbol{v}$. In order to analyze the behavior of different decision rules in this scenario, we redefine the H_1 hypothesis as follows

$$H_1 : \begin{cases} \boldsymbol{r} = \alpha \boldsymbol{p} + \boldsymbol{n}, \\ \boldsymbol{r}_k = \boldsymbol{n}_k, \qquad k = 1, \ldots, K. \end{cases} \tag{3.1}$$

As to P_d, it will denote the probability to decide for H_1 when it is actually in force.

As already mentioned, existing detectors behave quite differently when $\boldsymbol{p}$ is not aligned with $\boldsymbol{v}$ and can be classified as *robust*, *selective*, and *tunable* receivers. Obviously, it is desirable that P_d falls off rapidly as the degree of mismatch between the received signal and the postulated one increases (namely to achieve a small probability of sidelobe detection). At the same time, it is also necessary not to compromise the detection of slightly mismatched signals (mainlobe targets). Since previous requirement cannot be attained at the same time, it becomes important to trade good detection performance of mainlobe signals for rejection capabilities of sidelobe ones.

The remainder of this chapter focuses on several adaptive detection strategies proposed in the open literature. As a preliminary step, it is convenient to define a suitable measure for the mismatch in the *whitened* observation space, i.e., after whitening the nominal and the actual steering vectors with the true noise covariance matrix [37, 38]. To this end, we apply the transformation $\boldsymbol{M}^{-1/2}$ to the vector $\boldsymbol{p}$ and "rotate" the transformed vector as $\boldsymbol{U}\boldsymbol{M}^{-1/2}\boldsymbol{p}$, where $\boldsymbol{U}$ is a unitary matrix that rotates $\boldsymbol{M}^{-1/2}\boldsymbol{v}$ onto the vector $\boldsymbol{e}_1$ of the standard basis[1] of $\mathbb{C}^{N\times 1}$, i.e.,

$$\boldsymbol{U}\boldsymbol{M}^{-1/2}\boldsymbol{v} = \sqrt{\boldsymbol{v}^\dagger \boldsymbol{M}^{-1}\boldsymbol{v}}\, \boldsymbol{e}_1. \tag{3.2}$$

Denoting by $\boldsymbol{t}$ the unit vector aligned with $\boldsymbol{U}\boldsymbol{M}^{-1/2}\boldsymbol{p}$ we have that

$$\boldsymbol{U}\boldsymbol{M}^{-1/2}\boldsymbol{p} = \sqrt{\boldsymbol{p}^\dagger \boldsymbol{M}^{-1}\boldsymbol{p}}\, \frac{\boldsymbol{U}\boldsymbol{M}^{-1/2}\boldsymbol{p}}{\sqrt{\boldsymbol{p}^\dagger \boldsymbol{M}^{-1}\boldsymbol{p}}} = a_p \boldsymbol{t}\,, \tag{3.3}$$

with $a_p = \sqrt{\boldsymbol{p}^\dagger \boldsymbol{M}^{-1}\boldsymbol{p}}$. Then, we decompose the unit vector $\boldsymbol{t}$ into a component parallel to $\boldsymbol{e}_1$ and an orthogonal remainder

$$\boldsymbol{t} = \boldsymbol{t}^{\parallel} + \boldsymbol{t}^{\perp} = (\boldsymbol{e}_1^\dagger \boldsymbol{t})\boldsymbol{e}_1 + [\boldsymbol{t} - (\boldsymbol{e}_1^\dagger \boldsymbol{t})\boldsymbol{e}_1]. \tag{3.4}$$

The parallel component is given by

$$\boldsymbol{t}^{\parallel} = \left(\boldsymbol{e}_1^\dagger \frac{\boldsymbol{U}\boldsymbol{M}^{-1/2}\boldsymbol{p}}{a_p}\right)\boldsymbol{e}_1 = \frac{\boldsymbol{v}^\dagger \boldsymbol{M}^{-1}\boldsymbol{p}}{\sqrt{\boldsymbol{p}^\dagger \boldsymbol{M}^{-1}\boldsymbol{p}}\sqrt{\boldsymbol{v}^\dagger \boldsymbol{M}^{-1}\boldsymbol{v}}}\boldsymbol{e}_1. \tag{3.5}$$

[1] $\boldsymbol{e}_i, i = 1, \ldots, N$, has a 1 as its i-th component and 0's elsewhere.

Since the term $\boldsymbol{v}^\dagger \boldsymbol{M}^{-1} \boldsymbol{p}$ can be interpreted as an inner product of $\boldsymbol{v}$ and $\boldsymbol{p}$, it is possible to define a mismatch angle as follows

$$\frac{\boldsymbol{v}^\dagger \boldsymbol{M}^{-1} \boldsymbol{p}}{\sqrt{\boldsymbol{p}^\dagger \boldsymbol{M}^{-1} \boldsymbol{p}}\sqrt{\boldsymbol{v}^\dagger \boldsymbol{M}^{-1} \boldsymbol{v}}} = \frac{|\boldsymbol{v}^\dagger \boldsymbol{M}^{-1} \boldsymbol{p}|}{\sqrt{\boldsymbol{p}^\dagger \boldsymbol{M}^{-1} \boldsymbol{p}}\sqrt{\boldsymbol{v}^\dagger \boldsymbol{M}^{-1} \boldsymbol{v}}} e^{j\xi} = e^{j\xi} \cos\theta, \tag{3.6}$$

where $\xi \in [0, 2\pi)$ is the phase of $\boldsymbol{v}^\dagger \boldsymbol{M}^{-1} \boldsymbol{p}$ and $\theta \in [0, \pi/2]$ is the mismatch angle between the actual steering vector and the nominal one in the whitened observation space. Substituting (3.6) in (3.5) yields

$$\boldsymbol{t}^\parallel = e^{j\xi} \cos\theta \, \boldsymbol{e}_1. \tag{3.7}$$

Finally, since the Euclidean norm of the orthogonal component $\boldsymbol{t}^\perp$ is

$$\|\boldsymbol{t}^\perp\| = \sqrt{[\boldsymbol{t} - (\boldsymbol{e}_1^\dagger \boldsymbol{t})\boldsymbol{e}_1]^\dagger [\boldsymbol{t} - (\boldsymbol{e}_1^\dagger \boldsymbol{t})\boldsymbol{e}_1]} = \sqrt{1 - \cos^2\theta} = \sin\theta,$$

$\boldsymbol{t}$ can be recast as

$$\boldsymbol{t} = e^{j\xi} \cos\theta \, \boldsymbol{e}_1 + \sin\theta \begin{bmatrix} 0 \\ \boldsymbol{h} \end{bmatrix} = \begin{bmatrix} e^{j\xi} \cos\theta \\ \boldsymbol{h} \sin\theta \end{bmatrix}, \tag{3.8}$$

where $\boldsymbol{h} \in \mathbb{C}^{(N-1)\times 1}$ is a unit vector too. Remarkably, P_d of Kelly's detector, AMF, and ACE depends on the total available SNR, i.e.,

$$\mathrm{SNR} = |\alpha|^2 \boldsymbol{p}^\dagger \boldsymbol{M}^{-1} \boldsymbol{p}, \tag{3.9}$$

and $\cos^2\theta$ (in addition to the threshold) as shown in Appendix B. In the following the obvious dependence of P_d of any detector on the threshold will be omitted.

3.2 ROBUST RECEIVERS

Receivers belonging to this class provide good detection performance in presence of sensibly mismatched signals. A well-known robust receiver is the AMF, given by equation (2.23), and derived by means of the two-step GLRT-based design procedure for problem (2.11). Its performance for matched signals is similar to that of Kelly's detector, given by equation (2.21). However, in case of mismatched signals, the AMF exhibits a high probability of sidelobe detection as shown in Figure 3.2 where we plot P_d versus (the total available) SNR, defined by equation (3.9), assuming $N = 16$, $K = 32$, $P_{fa} = 10^{-4}$, and different values of $\cos^2\theta$.

The mismatched signal detection performance of a receiver can also be analyzed inspecting a 2D-graph, wherein the contours of constant P_d are represented as a function of the squared cosine of the mismatch angle, plotted vertically, and the SNR, plotted horizontally. Such plots were introduced in [4], where they are referred to as *mesa plots*.

In Figure 3.3 we plot contours of constant P_d for the AMF considering the same parameter setting of Figure 3.2. The behavior of the mesa for low $\cos^2\theta$ values shows that the AMF is a robust receiver, but also that it suffers the presence of *floors*, i.e., it can detect a target with $P_d \neq 0$ even

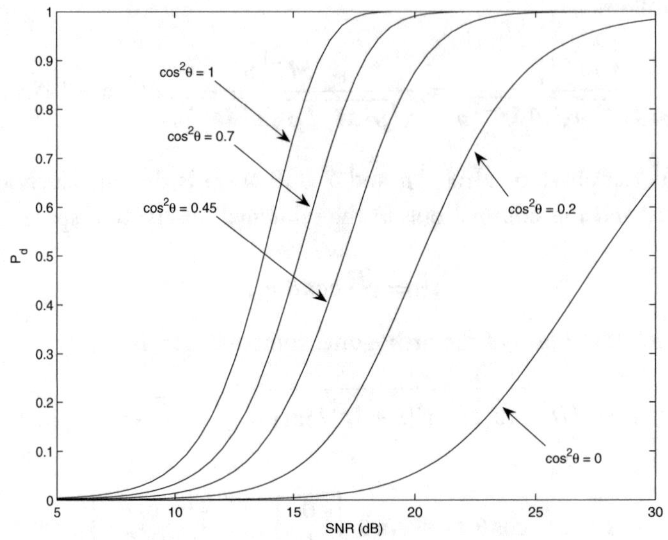

Figure 3.2: P_d vs SNR for the AMF with $N = 16$, $K = 32$, $P_{fa} = 10^{-4}$, and different values of $\cos^2\theta$.

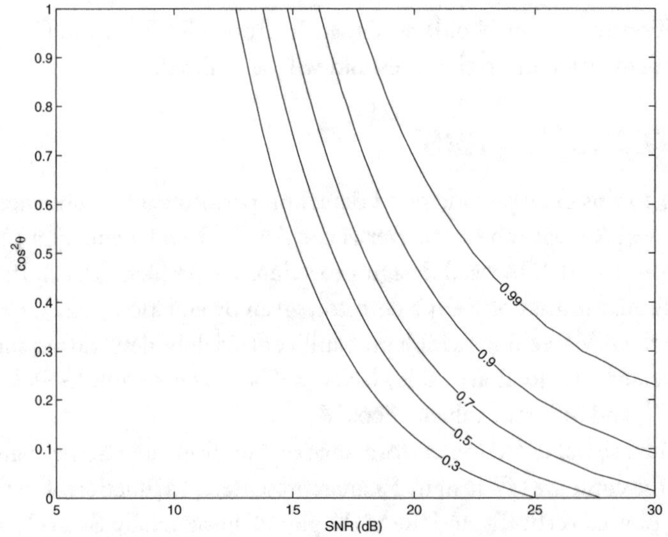

Figure 3.3: Contours of constant P_d for the AMF with $N = 16$, $K = 32$, and $P_{fa} = 10^{-4}$.

if $\cos^2\theta = 0$. Another way to visualize the presence of floors is to set $N_p = 1$ and plot P_d versus the difference between the nominal and the actual angle of arrival of the impinging signal echo, ψ_T say, as shown in Figure 3.4. Note though that plotting the P_d of the AMF versus ψ_T requires

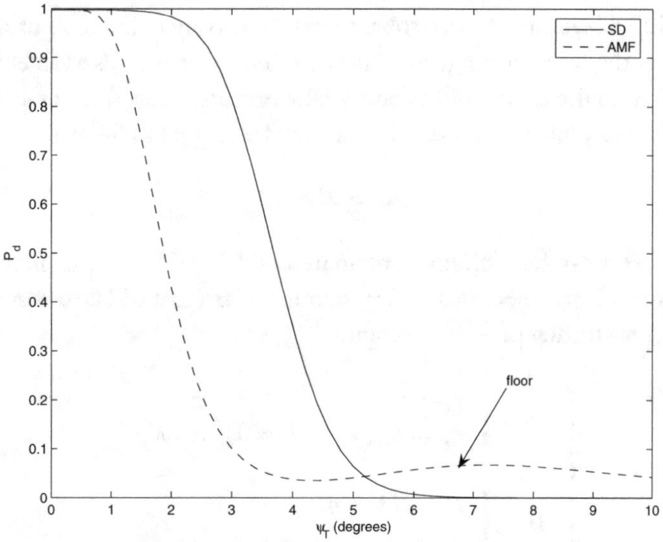

Figure 3.4: P_d vs ψ_T for the AMF and the SD with $N = N_a = 16$, $K = 32$, $r = 2$, $P_{fa} = 10^{-4}$, and SNR=19 dB.

specifying $\boldsymbol{v}$, $\boldsymbol{p}$, and $\boldsymbol{M}$. As to $\boldsymbol{v}$ ($\boldsymbol{p}$), we assume the expression of the target steering vector given by equation (2.10) with $d = \lambda/2$ and $\psi = \pi/2$ ($\psi = \pi/2 - \psi_T$). For future convenience, but with a little abuse of notation, we also rewrite equation (2.10) as follows

$$\boldsymbol{a}(\psi) = \begin{bmatrix} 1 \\ e^{j\pi\cos\psi} \\ \vdots \\ e^{j(N-1)\pi\cos\psi} \end{bmatrix}. \tag{3.10}$$

Accordingly, the nominal steering vector is

$$\boldsymbol{v} = \boldsymbol{a}(\pi/2), \tag{3.11}$$

while the actual steering vector is

$$\boldsymbol{p} = \boldsymbol{a}(\pi/2 - \psi_T). \tag{3.12}$$

As to the disturbance vectors, we resort to exponentially-correlated complex normal random vectors with one-lag correlation coefficient $\rho = 0.95$, namely the (i, j)-th element, $i, j = 1, \ldots, N$, of the covariance matrix $\boldsymbol{M}$ is given by $\rho^{|i-j|}$. These parameters will be used whenever it is necessary (unless otherwise specified). Figure 3.4 also assumes $N = N_a = 16$, $K = 32$, $P_{fa} = 10^{-4}$, and SNR= 19 dB.

It is possible to achieve an increased robustness by resorting to the tools of subspace detection, namely assuming that the target belongs to a known subspace of the observables [8, 9]. A subspace target model may capture the energy of the potentially distorted wavefront of a mainlobe target. In other words, it can be convenient to model the mismatched signal as follows

$$\boldsymbol{p} = \boldsymbol{H}\boldsymbol{b},$$

where $\boldsymbol{H} \in \mathbb{C}^{N\times r}$ is a known full-column-rank matrix and $\boldsymbol{b} \in \mathbb{C}^{r\times 1}$ is the unknown vector of the coordinates of $\boldsymbol{p}$ in the subspace spanned by the columns of $\boldsymbol{H}$ (that will be denoted in the following by $\langle \boldsymbol{H} \rangle$). The hypothesis testing problem becomes

$$\begin{cases} H_0 : \begin{cases} \boldsymbol{r} = \boldsymbol{n}, \\ \boldsymbol{r}_k = \boldsymbol{n}_k, \qquad k = 1, \ldots, K, \end{cases} \\ \\ H_1 : \begin{cases} \boldsymbol{r} = \boldsymbol{H}\boldsymbol{b} + \boldsymbol{n}, \\ \boldsymbol{r}_k = \boldsymbol{n}_k, \qquad k = 1, \ldots, K. \end{cases} \end{cases}$$

In [9] it is shown that the subspace detector (SD) implementing the GLRT for the above problem is

$$t_{\mathrm{SD}} = \frac{\boldsymbol{r}^\dagger \boldsymbol{S}^{-1}\boldsymbol{H}(\boldsymbol{H}^\dagger \boldsymbol{S}^{-1}\boldsymbol{H})^{-1}\boldsymbol{H}^\dagger \boldsymbol{S}^{-1}\boldsymbol{r}}{1 + \boldsymbol{r}^\dagger \boldsymbol{S}^{-1}\boldsymbol{r}} \underset{H_0}{\overset{H_1}{\gtrless}} \eta; \tag{3.13}$$

such result generalizes (to the multirank signal model) Kelly's detector (2.21).

Appendix C contains the statistical characterization of the SD (see [9] for further details); it turns out that the P_{fa} is independent of $\boldsymbol{M}$ and $\boldsymbol{H}$ (but for the value of r). As to the expression of P_d for matched signals, it depends on the SNR and r (it is otherwise independent of $\boldsymbol{H}$ provided that $\langle \boldsymbol{H} \rangle$ contains the nominal steering vector $\boldsymbol{v}$). Hereafter we assume that $\boldsymbol{v} \in \langle \boldsymbol{H} \rangle$ and, without loss of generality, that $\boldsymbol{v}$ is the first column of $\boldsymbol{H}$.

Specifying P_d for mismatched signals requires decomposing $\boldsymbol{M}^{-1/2}\boldsymbol{p}$ in terms of the orthogonal projection onto $\boldsymbol{M}^{-1/2}\boldsymbol{v}$ and the orthogonal projection onto the space spanned by the remaining columns of $\boldsymbol{H}_w = \boldsymbol{M}^{-1/2}\boldsymbol{H}$. As a preliminary step, we represent $\boldsymbol{H}_w$ in terms of the following QR factorization

$$\boldsymbol{H}_w = \boldsymbol{H}_0 \boldsymbol{T}_H,$$

where $\boldsymbol{H}_0 \in \mathbb{C}^{N\times r}$ is a slice of unitary matrix, namely $\boldsymbol{H}_0^\dagger \boldsymbol{H}_0 = \boldsymbol{I}$, and $\boldsymbol{T}_H \in \mathbb{C}^{r\times r}$ is an invertible upper triangular matrix. Then, among the unitary matrices $\boldsymbol{U}$ that rotate $\boldsymbol{M}^{-1/2}\boldsymbol{v}$ onto $\boldsymbol{e}_1$ (see equation (3.2)), we choose one that rotates the r orthonormal columns of $\boldsymbol{H}_0$ onto the first r vectors of the standard basis of $\mathbb{C}^{N\times 1}$, i.e.,

$$\boldsymbol{U}\boldsymbol{H}_0 = \boldsymbol{E}_r = [\boldsymbol{e}_1 \cdots \boldsymbol{e}_r] = \begin{bmatrix} \boldsymbol{I} \\ \boldsymbol{0} \end{bmatrix}.$$

Decomposing the unit vector $\boldsymbol{h}$ of equation (3.8) as

$$\boldsymbol{h} = \begin{bmatrix} \boldsymbol{h}_B \\ \boldsymbol{h}_C \end{bmatrix},$$

with $\boldsymbol{h}_B \in \mathbb{C}^{(r-1)\times 1}$ and $\boldsymbol{h}_C \in \mathbb{C}^{(N-r)\times 1}$, the unit vector $\boldsymbol{t}$, given by (3.8), becomes[2]

$$\boldsymbol{t} = \begin{bmatrix} e^{j\xi}\cos\theta \\ \boldsymbol{h}_B \sin\theta \\ \boldsymbol{h}_C \sin\theta \end{bmatrix}. \tag{3.14}$$

Remarkably, Appendix C shows that P_d for mismatched signals depends on SNR, $\cos^2\theta$, and $\|\boldsymbol{h}_B\|$ (or $\|\boldsymbol{h}_C\|$).

In Figures 3.4 and 3.5 we evaluate the robustness of the SD with respect to mismatched

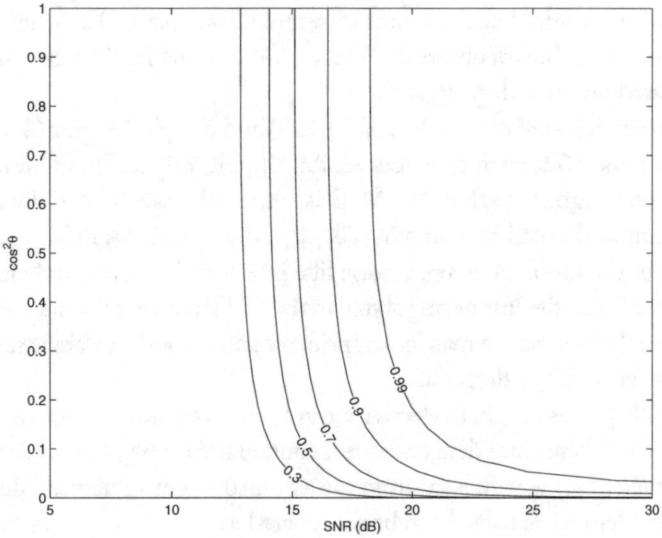

Figure 3.5: Contours of constant P_d for the SD with $N = N_a = 16$, $K = 32$, $r = 2$, and $P_{fa} = 10^{-4}$.

signals assuming $N = N_a = 16$, $K = 32$, $r = 2$, $P_{fa} = 10^{-4}$ (and the nominal angle of arrival

[2]Note that $\boldsymbol{h}_B$ and $\boldsymbol{h}_C$ are such that $\|\boldsymbol{h}_B\|^2 + \|\boldsymbol{h}_C\|^2 = 1$.

equal to $\pi/2$). In addition, the matrix $\boldsymbol{H}$ is chosen as

$$\boldsymbol{H} = [\boldsymbol{v}\ \boldsymbol{v}_1],$$

where $\boldsymbol{v}_1$ is a slightly mismatched steering vector (corresponding to an angle of arrival of $\pi/2 - \pi/360$) which returns

$$\frac{|\boldsymbol{v}_1^\dagger \boldsymbol{M}^{-1}\boldsymbol{v}|^2}{(\boldsymbol{v}^\dagger \boldsymbol{M}^{-1}\boldsymbol{v})(\boldsymbol{v}_1^\dagger \boldsymbol{M}^{-1}\boldsymbol{v}_1)} = 0.895.$$

The figures show that the SD can guarantee a superior robustness with respect to AMF; Figure 3.4 also highlights that the SD does not present any floor (for the considered parameter values). In Chapter 4 we will also discuss how to choose $\boldsymbol{H}$ and r in order to tune the robustness of two-stage detectors employing the SD.

3.3 SELECTIVE RECEIVERS

An important issue in radar detection is the capability to reject signals whose signature is unlikely to correspond to the signal of interest, in order not to generate false alarms.

The ACE, given by (2.22), provides excellent sidelobe rejection capabilities, at the price of a lower detection performance for matched signals with respect to Kelly's detector. It measures the squared cosine of the angle that the vector of returns from the CUT forms with the nominal steering vector in the *quasi whitened* observation space, i.e., after whitening the data with the sample covariance matrix based on secondary data.

Note that the decision statistics of the ACE (2.22) and Kelly's detector (2.21) are very similar, but, as it is shown in Figure 3.6, such detectors exhibit significantly different performances; indeed, inspection of the figure highlights that the ACE is more selective than Kelly's detector [39]. A theoretical justification of this can be found in [39, 40]. More precisely, in [40] it is shown that the ACE is the GLRT for the problem where a noise like interferer is present only in the CUT. When in addition it is assumed that the interfering signal in the CUT is such that the so-called generalized eigenrelation between the covariance matrices of primary and secondary data is satisfied, the GLRT turns out to coincide with Kelly's detector.

Detectors which possess higher selectivity can be derived modifying the conventional null hypothesis, which usually states that data under test contain noise only, so that data possibly contain a fictitious signal which, in some way, is far from the assumed target signature. More specifically, the hypothesis testing problem to be solved can be formulated as

$$\begin{cases} H_0 : \begin{cases} \boldsymbol{r} = \boldsymbol{v}_\perp + \boldsymbol{n}, & \\ \boldsymbol{r}_k = \boldsymbol{n}_k, & k = 1, \dots, K, \end{cases} \\ \\ H_1 : \begin{cases} \boldsymbol{r} = \alpha\boldsymbol{v} + \boldsymbol{n}, & \\ \boldsymbol{r}_k = \boldsymbol{n}_k, & k = 1, \dots, K, \end{cases} \end{cases}$$

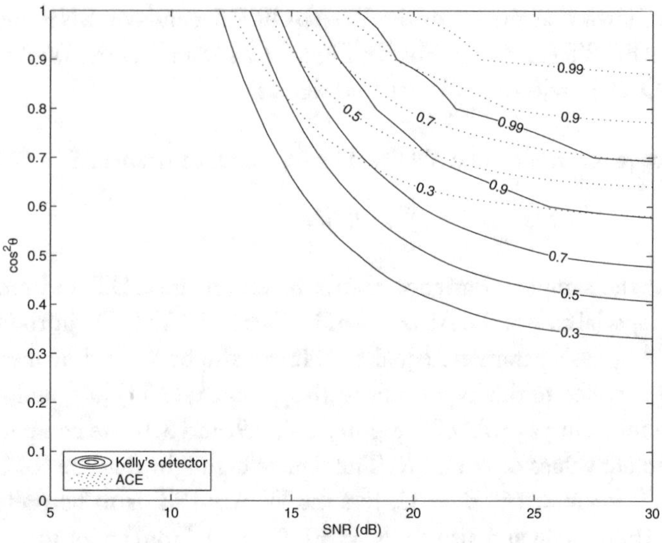

Figure 3.6: Contours of constant P_d for the ACE and Kelly's detector with $N = 16$, $K = 32$, and $P_{fa} = 10^{-4}$.

where $\boldsymbol{v}_\perp$ is the fictitious signal. Doing so, if a mismatched signal is present, the detector will be less inclined to declare a detection, as the null hypothesis will be more plausible than in the case where, under the null hypothesis, the test vector contains noise only. This is the essence of the adaptive beamformer orthogonal rejection test (ABORT) proposed in [4], where $\boldsymbol{v}_\perp$ is assumed to be orthogonal to $\boldsymbol{v}$ in the quasi-whitened space. For reader's ease we recall that the decision statistic of the ABORT is given by

$$t_{\text{A}} = \frac{1 + \dfrac{\left|\boldsymbol{r}^\dagger \boldsymbol{S}^{-1}\boldsymbol{v}\right|^2}{\boldsymbol{v}^\dagger \boldsymbol{S}^{-1}\boldsymbol{v}}}{2 + \boldsymbol{r}^\dagger \boldsymbol{S}^{-1}\boldsymbol{r}} = \frac{1 + t_{\text{AMF}}}{2 + \boldsymbol{r}^\dagger \boldsymbol{S}^{-1}\boldsymbol{r}}. \tag{3.15}$$

A GLRT-based modification of the ABORT has been proposed in [7], wherein the useful and the fictitious signals are orthogonal in the truly whitened observation space, i.e., after whitening with the true noise covariance matrix. This seemingly minor modification leads to significant differences compared to the ABORT and, more particularly, to enhanced rejection capabilities of sidelobe signals (although at the price of an additional loss with respect to Kelly's detector). The decision statistic of this detector, referred to in the following as W-ABORT, is given by

$$t_{\text{WA}} = \frac{1}{(1 + \boldsymbol{r}^\dagger \boldsymbol{S}^{-1}\boldsymbol{r})\left[1 - \dfrac{|\boldsymbol{r}^\dagger \boldsymbol{S}^{-1}\boldsymbol{v}|^2}{(\boldsymbol{v}^\dagger \boldsymbol{S}^{-1}\boldsymbol{v})(1 + \boldsymbol{r}^\dagger \boldsymbol{S}^{-1}\boldsymbol{r})}\right]^2}. \tag{3.16}$$

In Appendix B it is shown that the P_d of the W-ABORT depends on SNR and $\cos^2\theta$ only. The same is true for the ABORT (3.15) [4]. Finally, Figure 3.7 shows that the W-ABORT is definitely superior to the ABORT in rejecting mismatched signals[3].

Another selective receiver can be obtained by replacing the matrix $\boldsymbol{S}$ in (2.23) with

$$\overline{\boldsymbol{S}} = \boldsymbol{S} + \boldsymbol{r}\boldsymbol{r}^{\dagger},$$

namely $K+1$ times the sample covariance matrix based on the CUT and secondary data. The resulting receiver is a special case of the AMF *with De-Emphasis* (AMFD), introduced by Richmond in [41], with the de-emphasis parameter equal to 1. It can also be derived by means of the Rao test design criterion [28], applied to the hypothesis testing problem (2.11) [18] and for this reason will be referred to in the following as RAO. In Figures 3.8, 3.9, and 3.10 we compare the RAO to the W-ABORT for different values of N and K. The figures highlight that the RAO is more selective than the W-ABORT for $N = 16$, $K = 32$, but the W-ABORT is to be preferred (in terms of selectivity) for $N = 16$, $K = 48$ and also for $N = 30$, $K = 60$. From an intuitive point of view, this result is not surprising since if $K \to +\infty$ (for a given SNR) the contribution of the primary data to the sample covariance becomes negligible, so making the RAO more similar to the AMF.

In Table 3.1, we present a ranking of the detection algorithms analyzed up to this point in terms of their robustness/selectivity.

Table 3.1: A comparison of existing detectors in terms of directivity.

Detector	Behavior
W-ABORT	very selective
RAO	very selective
ACE	selective
ABORT	selective
Kelly's detector	selective
AMF	robust
SD	very robust

3.4 TUNABLE RECEIVERS

The above analysis points out that it is difficult to find a decision scheme capable of providing at the same time good capabilities to reject sidelobe signals and high power in case of mainlobe targets. In

[3] It is important to stress that also for decision schemes derived assuming the presence of a fictitious signal, under the H_0 hypothesis, the P_{fa} is defined as the probability to choose H_1 under the noise-only hypothesis.

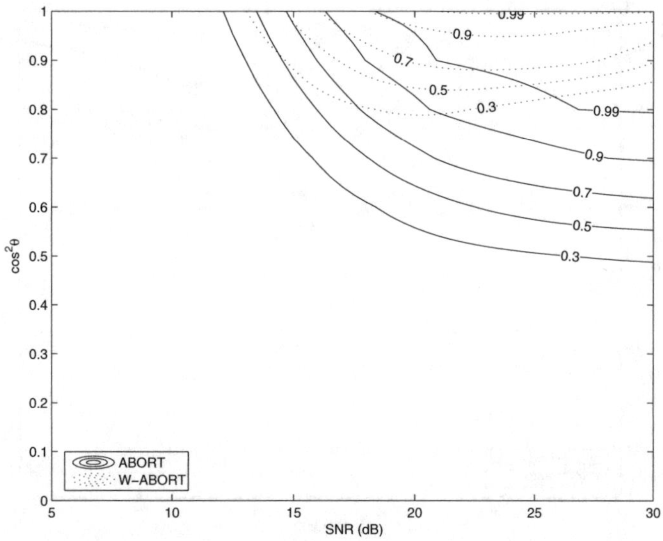

Figure 3.7: Contours of constant P_d for the ABORT and the W-ABORT with $N = 16$, $K = 32$, and $P_{fa} = 10^{-4}$.

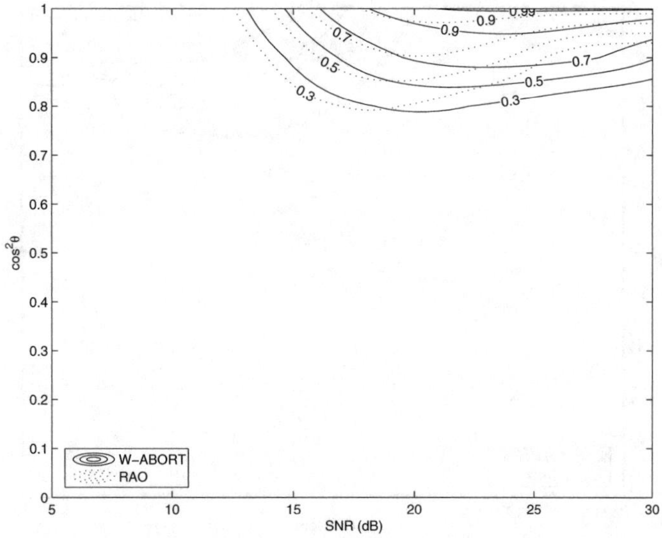

Figure 3.8: Contours of constant P_d for the W-ABORT and the RAO with $N = 16$, $K = 32$, and $P_{fa} = 10^{-4}$.

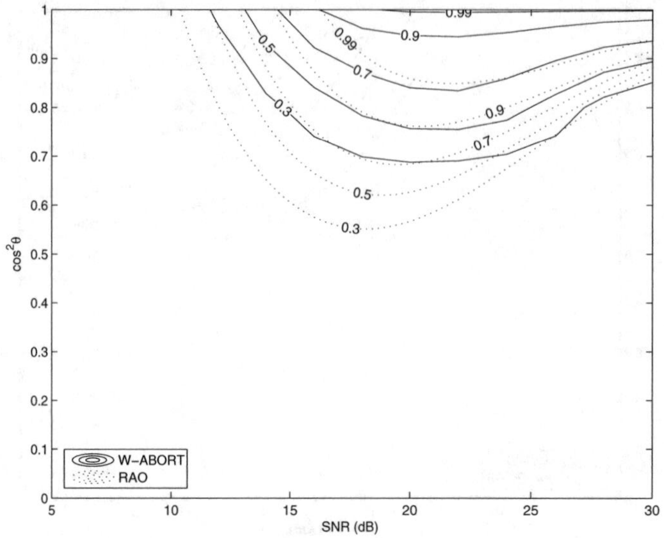

Figure 3.9: Contours of constant P_d for the W-ABORT and the RAO with $N = 16$, $K = 48$, and $P_{fa} = 10^{-4}$.

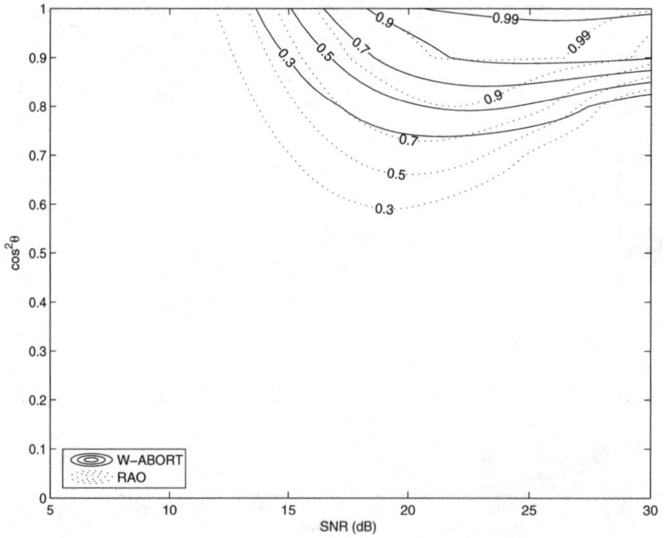

Figure 3.10: Contours of constant P_d for the W-ABORT and the RAO with $N = 30$, $K = 60$, and $P_{fa} = 10^{-4}$.

addition, it would be highly desirable to control the rate at which the probability of detection falls down when the actual steering vector departs from the nominal one. In order to accomplish this task, tunable receivers have been proposed. The term tunable concerns the fact that such detectors require to set some design parameter in order to obtain the desired behavior in terms of directivity. More precisely, the directivity of existing tunable receivers can be adjusted either by setting a parameter which appears in the decision statistic (the so-called *parametric receivers*) or by proper threshold setting (the so-called *two-stage detectors*). Examples of both parametric and two-stage receivers will be given below.

3.4.1 PARAMETRIC RECEIVERS

Such schemes can be obtained by exploiting the similarities existing between decision statistics of detectors with different behaviors. In particular, it is possible to manipulate those factors which make different the corresponding statistics by introducing a proper parameter in order to generate receiver operating characteristics in between those of the considered detectors.

For instance, observe that the difference between the decision statistic of the AMF (2.23) and that of Kelly's detector (2.21) is the term

$$1 + \boldsymbol{r}^{\dagger}\boldsymbol{S}^{-1}\boldsymbol{r},$$

which appears in the denominator of (2.21). Thus, the above detectors can be mixed as follows [10]

$$t = \frac{|\boldsymbol{r}^{\dagger}\boldsymbol{S}^{-1}\boldsymbol{v}|^2}{(\boldsymbol{v}^{\dagger}\boldsymbol{S}^{-1}\boldsymbol{v})(1 + \mu\boldsymbol{r}^{\dagger}\boldsymbol{S}^{-1}\boldsymbol{r})}, \tag{3.17}$$

where $\mu \in [0, 1]$. Note that by tuning the parameter μ, t can achieve the robustness of the AMF ($\mu = 0$) or the selectivity of Kelly's detector ($\mu = 1$), or an intermediate behavior.

Another parametric detector, aimed at increasing the selectivity of Kelly's detector, can be obtained by combining it with the W-ABORT [42]. More precisely, let us consider the following equivalent form of Kelly's detector

$$\frac{1}{1 - t_{\mathrm{K}}} = \frac{1 + \boldsymbol{r}^{\dagger}\boldsymbol{S}^{-1}\boldsymbol{r}}{1 + \boldsymbol{r}^{\dagger}\boldsymbol{S}^{-1}\boldsymbol{r} - \dfrac{|\boldsymbol{r}^{\dagger}\boldsymbol{S}^{-1}\boldsymbol{v}|^2}{\boldsymbol{v}^{\dagger}\boldsymbol{S}^{-1}\boldsymbol{v}}},$$

and rewrite equation (3.16) as

$$t_{\mathrm{WA}} = \frac{1 + \boldsymbol{r}^{\dagger}\boldsymbol{S}^{-1}\boldsymbol{r}}{\left[1 + \boldsymbol{r}^{\dagger}\boldsymbol{S}^{-1}\boldsymbol{r} - \dfrac{|\boldsymbol{r}^{\dagger}\boldsymbol{S}^{-1}\boldsymbol{v}|^2}{\boldsymbol{v}^{\dagger}\boldsymbol{S}^{-1}\boldsymbol{v}}\right]^2}. \tag{3.18}$$

Then, the parametric detector is given by

$$t_{\mathrm{KWA}} = \frac{1 + \boldsymbol{r}^{\dagger}\boldsymbol{S}^{-1}\boldsymbol{r}}{\left[1 + \boldsymbol{r}^{\dagger}\boldsymbol{S}^{-1}\boldsymbol{r} - \dfrac{|\boldsymbol{r}^{\dagger}\boldsymbol{S}^{-1}\boldsymbol{v}|^2}{\boldsymbol{v}^{\dagger}\boldsymbol{S}^{-1}\boldsymbol{v}}\right]^{2\mu}}, \tag{3.19}$$

where $\mu > 0$. Note that for $\mu = 1$ the above receiver corresponds to the W-ABORT, while for $\mu = 1/2$ it returns a decision statistic equivalent to Kelly's detector. Following a rationale similar to that used to derive detectors (3.17) and (3.19) other alternatives might be conceived.

Another family of parametric receivers can be constructed exploiting the so-called *conic acceptance idea* [11, 12], which is tantamount to constraining the possible useful signal to belong to a proper cone with axis the nominal steering vector in the whitened observation space. The hypothesis testing problem with a conic acceptance is

$$\begin{cases} H_0 : \begin{cases} \boldsymbol{r} = \boldsymbol{n}, & \\ \boldsymbol{r}_k = \boldsymbol{n}_k, & k = 1, \ldots, K, \end{cases} \\ \\ H_1 : \begin{cases} \boldsymbol{r} = \alpha \boldsymbol{p} + \boldsymbol{n}, & \boldsymbol{M}^{-1/2}\boldsymbol{p} \in \Gamma_{\epsilon}(\boldsymbol{M}^{-1/2}\boldsymbol{v}), \\ \boldsymbol{r}_k = \boldsymbol{n}_k, & k = 1, \ldots, K, \end{cases} \end{cases}$$

where $\boldsymbol{p}$ is the actual steering vector possibly different from the nominal one $\boldsymbol{v}$ and

$$\Gamma_{\epsilon}(\boldsymbol{s}) = \left\{ \boldsymbol{z} \in \mathbb{C}^{N \times 1} : \|\boldsymbol{z}\|^2 \le (1 + \epsilon^2) \frac{|\boldsymbol{z}^{\dagger}\boldsymbol{s}|^2}{\|\boldsymbol{s}\|^2} \right\}, \quad \epsilon > 0.$$

It is not difficult to show that the two-step GLRT-based design procedure for the above problem leads to the following test, referred to in the following as conic-acceptance detector (CAD), see also [12]

$$t_{\mathrm{CAD}} = \boldsymbol{r}^{\dagger}\boldsymbol{S}^{-1}\boldsymbol{r} - \frac{1}{1+\epsilon^2}\left[\sqrt{\boldsymbol{r}^{\dagger}\boldsymbol{S}^{-1}\boldsymbol{r} - \frac{|\boldsymbol{r}^{\dagger}\boldsymbol{S}^{-1}\boldsymbol{v}|^2}{\boldsymbol{v}^{\dagger}\boldsymbol{S}^{-1}\boldsymbol{v}}} - \epsilon\sqrt{\frac{|\boldsymbol{r}^{\dagger}\boldsymbol{S}^{-1}\boldsymbol{v}|^2}{\boldsymbol{v}^{\dagger}\boldsymbol{S}^{-1}\boldsymbol{v}}}\right]^2$$
$$\times\, u\left(\boldsymbol{r}^{\dagger}\boldsymbol{S}^{-1}\boldsymbol{r} - \frac{|\boldsymbol{r}^{\dagger}\boldsymbol{S}^{-1}\boldsymbol{v}|^2}{\boldsymbol{v}^{\dagger}\boldsymbol{S}^{-1}\boldsymbol{v}}(1+\epsilon^2)\right) \underset{H_0}{\overset{H_1}{\gtrless}} \eta,$$

where $u(x)$ is the unit-step function, i.e.,

$$u(x) = \begin{cases} 1, & x \ge 0, \\ 0, & x < 0. \end{cases}$$

It is also possible to prove, based on results contained in Appendix B, that P_d of the CAD depends on SNR, $\cos^2\theta$, and ϵ only. In particular, the parameter ϵ, which governs the aperture of the acceptance

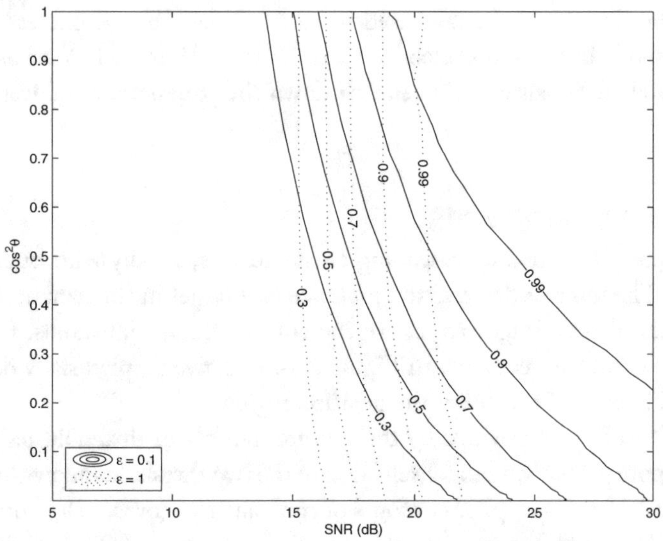

Figure 3.11: Contours of constant P_d for the CAD with $N = 8$, $K = 16$, and $P_{fa} = 10^{-4}$.

cone, allows to vary the robustness of the receiver to mismatched signals, as shown in Figure 3.11. However, the figure points out that the CAD is also characterized by very high floors. The limiting behaviors of the CAD are those of the AMF and of the energy detector, namely

$$\begin{aligned} \epsilon \to 0: &\quad t_{\text{CAD}} \to t_{\text{AMF}}, \\ \epsilon \to \infty: &\quad t_{\text{CAD}} \to \boldsymbol{r}^\dagger \boldsymbol{S}^{-1} \boldsymbol{r}. \end{aligned}$$

The cone idea has also been used assuming a signal that belongs to a cone with axis the nominal steering vector under H_1 and a fictitious signal in the complement of this cone under H_0 [12]. As a matter of fact, the two-step GLRT-based design procedure for this problem leads to the following conic-acceptance-rejection detector (CARD)

$$t_{\text{CARD}} = \left[\epsilon \sqrt{\frac{|\boldsymbol{r}^\dagger \boldsymbol{S}^{-1} \boldsymbol{v}|^2}{\boldsymbol{v}^\dagger \boldsymbol{S}^{-1} \boldsymbol{v}}} - \sqrt{\boldsymbol{r}^\dagger \boldsymbol{S}^{-1} \boldsymbol{r} - \frac{|\boldsymbol{r}^\dagger \boldsymbol{S}^{-1} \boldsymbol{v}|^2}{\boldsymbol{v}^\dagger \boldsymbol{S}^{-1} \boldsymbol{v}}} \right]^2 \times \operatorname{sign}\left(\epsilon \sqrt{\frac{|\boldsymbol{r}^\dagger \boldsymbol{S}^{-1} \boldsymbol{v}|^2}{\boldsymbol{v}^\dagger \boldsymbol{S}^{-1} \boldsymbol{v}}} - \sqrt{\boldsymbol{r}^\dagger \boldsymbol{S}^{-1} \boldsymbol{r} - \frac{|\boldsymbol{r}^\dagger \boldsymbol{S}^{-1} \boldsymbol{v}|^2}{\boldsymbol{v}^\dagger \boldsymbol{S}^{-1} \boldsymbol{v}}} \right) \underset{H_0}{\overset{H_1}{\gtrless}} \eta,$$

where $\operatorname{sign}(x)$ is the signum function, i.e.,

$$\operatorname{sign}(x) = \begin{cases} +1, & x \geq 0, \\ -1, & x < 0. \end{cases}$$

The selectivity of the CARD can be increased by tuning down the parameter[4] ϵ. However, Figures 3.12 and 3.13 show that it is less selective than the W-ABORT when we assume comparable performance for matched signals and that tuning down the parameter ϵ reduces the possibility to detect matched signals.

3.4.2 TWO-STAGE RECEIVERS

This class of receivers is obtained by cascading two detectors, usually with opposite behaviors in terms of directivity. The receiver declares the presence of a target in the data under test only when the decision statistics of both stages are above the corresponding thresholds. It can be seen as a logical AND between the two receivers. In Figure 3.14 it is given a pictorially description of such architecture, while Figure 3.15 illustrates the decision regions.

The tuning capability is provided by the infinite number of threshold pairs that ensure the same value of P_{fa}; more precisely, a proper selection of the two thresholds allows to adjust directivity. For example, Figure 3.16 shows typical contours of constant P_{fa} (or iso-P_{fa} contours) as function of the two thresholds. In order to increase the resulting robustness (or selectivity), it is necessary to choose threshold pairs moving toward the robust (or selective) stage on the iso-P_{fa} contour. In contrast to parametric receivers, two-stage detection strategies, due to their modular structure, typically guarantee a wider range of directivity.

A rather famous two-stage detector is the adaptive sidelobe blanker (ASB), which consists of the cascade of the AMF and the ACE. Such detector has been proposed as an effective means for mitigating the high number of false alarms of the AMF due to the presence of clutter inhomogeneities [43, 44, 45]. Note that the ASB encompasses as special cases the AMF and the ACE, in fact

$$\begin{array}{lclcl} \text{Threshold}_{\text{AMF}} > 0 & \text{and} & \text{Threshold}_{\text{ACE}} = 0 & \Rightarrow & \text{ASB} \equiv \text{AMF} \\ \text{Threshold}_{\text{AMF}} = 0 & \text{and} & \text{Threshold}_{\text{ACE}} > 0 & \Rightarrow & \text{ASB} \equiv \text{ACE} \end{array}$$

Obviously, it offers infinite gradations between the AMF and the ACE, so giving the possibility to trade good rejection capabilities of sidelobe signals for acceptable loss of matched signals [41, 46]. Richmond has also provided closed-form expressions for the P_d and P_{fa} of the ASB and demonstrated that in homogeneous environment and with matched signals it has higher or commensurate P_d for a given P_{fa} than both the AMF and the ACE and an overall performance that is commensurate with Kelly's detector [17]. In [46] a further two-stage detector consisting of the AMF followed by Kelly's detector has been proposed as a computationally efficient implementation of the latter. Finally, in [18] a two-stage detector obtained cascading a RAO test to the AMF has been proposed and assessed.

[4] Again, P_d depends on SNR, $\cos^2\theta$, and ϵ only.

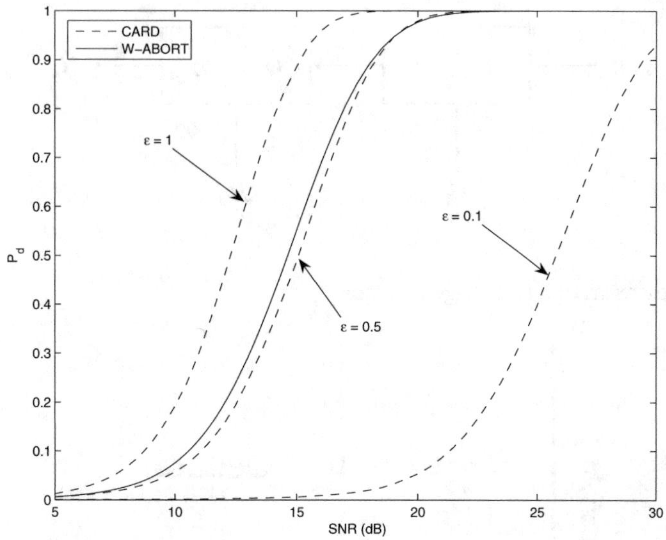

Figure 3.12: P_d versus SNR in case of matched signals for the W-ABORT and the CARD with $N = 16$, $K = 32$, and $P_{fa} = 10^{-4}$.

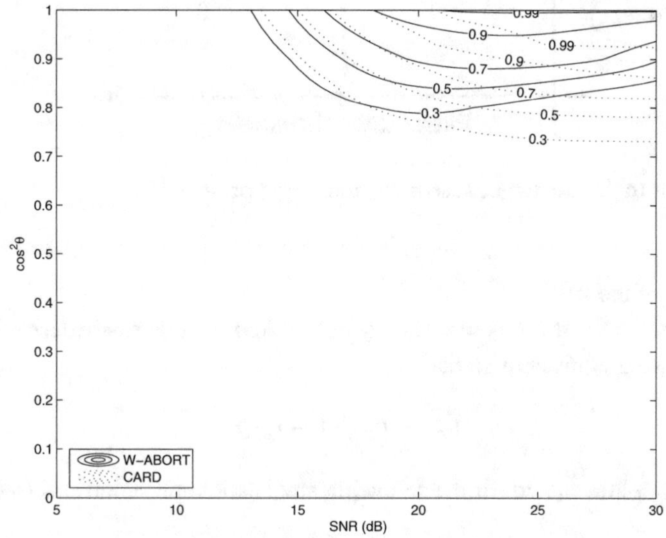

Figure 3.13: Contours of constant P_d for the W-ABORT and the CARD with $N = 16$, $K = 32$, $\epsilon = 0.5$, and $P_{fa} = 10^{-4}$.

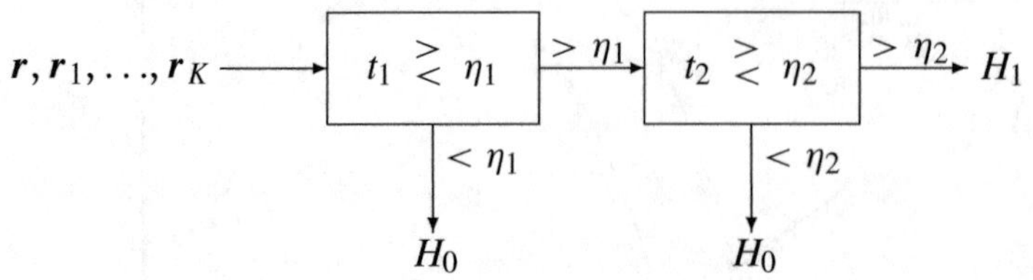

Figure 3.14: Block diagram of a two-stage detector.

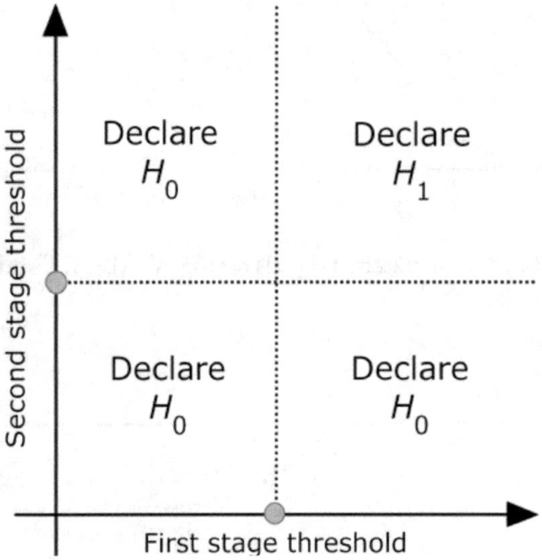

Figure 3.15: Regions to declare target absent H_0 and target present H_1.

3.4.2.1 P_{fa} and P_d of the ASB

Closed form expressions for the P_{fa} and the P_d of the ASB can be easily derived by replacing the ACE with the following equivalent statistic

$$\tilde{t}_{\mathrm{ACE}} = t_{\mathrm{ACE}}/(1 - t_{\mathrm{ACE}})$$

and observing that t_{AMF} and $\tilde{t}_{\mathrm{ACE}}$ admit the following stochastic representations (see Appendix B)

$$t_{\mathrm{AMF}} = \tilde{t}_{\mathrm{K}}/\beta \quad \text{and} \quad \tilde{t}_{\mathrm{ACE}} = \tilde{t}_{\mathrm{K}}/(1 - \beta),$$

where

$$\tilde{t}_{\mathrm{K}} = \frac{t_{\mathrm{K}}}{1 - t_{\mathrm{K}}},$$

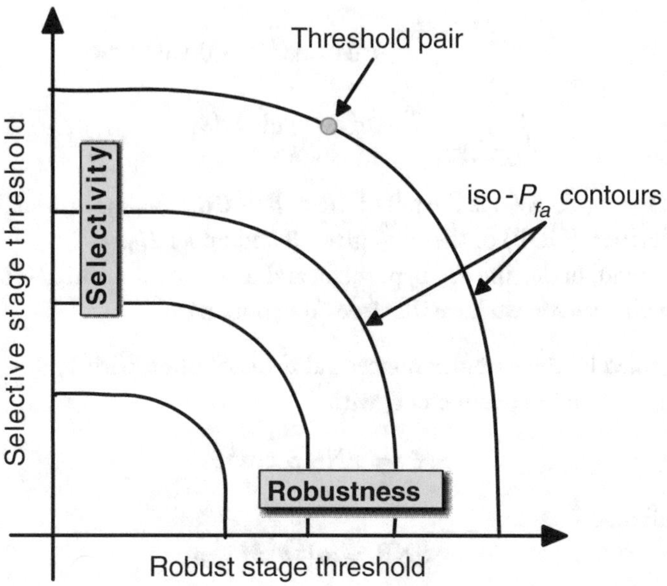

Figure 3.16: Contours of constant P_{fa} of a two-stage detector.

with t_K given by the left-hand side of (2.21).

Thus, under the H_0 hypothesis, it can be shown that (see Appendix B)

- $\tilde{t}_K$, given β, is ruled by a complex central F-distribution with $1, K-N+1$ degrees of freedom[5] [2];

- β is a complex central beta distributed rv with $K-N+2$, $N-1$ degrees of freedom, i.e., $\beta \sim \mathcal{C}\beta_{K-N+2,N-1}$.

It follows that the P_{fa} of the ASB can be computed as

$$\begin{aligned}
P_{fa}(\eta_{\text{AMF}}, \eta_{\text{ACE}}) &= \mathrm{P}[t_{\text{AMF}} > \eta_{\text{AMF}},\ \tilde{t}_{\text{ACE}} > \tilde{\eta}_{\text{ACE}};\ H_0] \\
&= \mathrm{P}\left[\frac{\tilde{t}_K}{\beta} > \eta_{\text{AMF}},\ \frac{\tilde{t}_K}{1-\beta} > \tilde{\eta}_{\text{ACE}}\right] \\
&= \int_0^1 \mathrm{P}\left[\tilde{t}_K > \eta_{\text{AMF}} x,\ \tilde{t}_K > \tilde{\eta}_{\text{ACE}}(1-x) \mid \beta = x; H_0\right] p_0(x)\, dx \\
&= 1 - \int_0^1 \mathcal{P}_0\left(\max(\eta_{\text{AMF}} x, \tilde{\eta}_{\text{ACE}}(1-x))\right) p_0(x)\, dx
\end{aligned}$$

[5]For a definition of complex normal related statistics see Appendix A.

$$= 1 - \int_0^{\frac{\tilde{\eta}_{ACE}}{\eta_{AMF}+\tilde{\eta}_{ACE}}} \mathcal{P}_0\left(\tilde{\eta}_{ACE}(1-x)\right) p_0(x)\, dx - \int_{\frac{\tilde{\eta}_{ACE}}{\eta_{AMF}+\tilde{\eta}_{ACE}}}^1 \mathcal{P}_0\left(\eta_{AMF} x\right) p_0(x)\, dx,$$

where $\tilde{\eta}_{ACE} = \eta_{ACE}/(1-\eta_{ACE})$, $p_0(\cdot)$ is the pdf of the rv $\beta \sim \mathcal{C}\beta_{K-N+2,N-1}$, and $\mathcal{P}_0(\cdot)$ is the cumulative distribution function (CDF) of the rv $\tilde{t}_K$, given β and under H_0.

On the other hand, under the H_1 hypothesis and assuming a mismatch between the actual and the nominal steering vector, we have that (see Appendix B)

- $\tilde{t}_K$, given β, is ruled by the complex noncentral F-distribution with $1, K-N+1$ degrees of freedom and noncentrality parameter δ, with

  $$\delta^2 = \text{SNR}\, \beta \cos^2\theta,$$

  where we recall that

  $$\text{SNR} = |\alpha|^2 \boldsymbol{p}^\dagger \boldsymbol{M}^{-1} \boldsymbol{p}$$

  is the total available signal-to-noise ratio;

- β is ruled by the complex noncentral beta distribution with $K-N+2, N-1$ degrees of freedom and noncentrality parameter δ_β, with

  $$\delta_\beta^2 = \text{SNR} \sin^2\theta,$$

  i.e., $\beta \sim \mathcal{C}\beta_{K-N+2,N-1}(\delta_\beta)$.

Thus, the P_d of the ASB is given by

$$P_d(\eta_{AMF}, \eta_{ACE}, \text{SNR}, \cos^2\theta) = 1 - \int_0^{\frac{\tilde{\eta}_{ACE}}{\eta_{AMF}+\tilde{\eta}_{ACE}}} \mathcal{P}_1\left(\tilde{\eta}_{ACE}(1-x)\right) p_1(x)\, dx - \int_{\frac{\tilde{\eta}_{ACE}}{\eta_{AMF}+\tilde{\eta}_{ACE}}}^1 \mathcal{P}_1\left(\eta_{AMF} x\right) p_1(x)\, dx,$$

where $p_1(\cdot)$ is the pdf of the rv $\beta \sim \mathcal{C}\beta_{K-N+2,N-1}(\delta_\beta)$ and $\mathcal{P}_1(\cdot)$ is the CDF of the rv $\tilde{t}_K$, given β and under H_1. In the evaluation of the above integral it is important to take into account that $\mathcal{P}_1(\cdot)$ depends on x also through $\delta^2 = \text{SNR}\, x \cos^2\theta$; this dependence has not been highlighted in order not to burden too much the formula.

The ASB (and the AMF-RAO) will be assessed in the next chapter together with other two-stage detectors which offer an extended operational range in terms of directivity with respect to the ASB.

CHAPTER 4

Enhanced Adaptive Sidelobe Blanking Algorithms

In Chapter 3 we have reviewed different classes of detectors. We have shown that the SD is capable of coping with highly mismatched signals and, more important, it does not suffer high floors, which are typical of AMF and CAD. On the other hand, the W-ABORT and the RAO detector exhibit a P_d which rapidly drops to zero as the degree of mismatch between the actual and the nominal steering vectors increases, with the difference that the RAO becomes more similar to the AMF as K increases (given the SNR). In addition, it can be shown that the ASB, obtained by cascading two detectors, with opposite behaviors in terms of selectivity, can trade "near optimal" detection capabilities of slightly mismatched signals for better rejection of sidelobe signals. Based upon the above considerations, in this chapter, we present other two-stage detectors, which provide a wider range of directivity values with respect to the ASB [13, 14, 18]. Since enhanced selectivity is typically paid in terms of a loss with respect to Kelly's detector that could exceed 1 dB, comparisons in terms of selectivity refer to threshold pairs that guarantee a loss of 1 dB at most with respect to Kelly's detector at $P_d = 0.9$ and $P_{fa} = 10^{-4}$.

The reminder of the chapter is organized as follows: in Section 4.1 we describe and analyze a two-stage decision scheme aimed at increasing the robustness of the ASB, while in Section 4.2 we introduce two-stage detectors which outperform the previous one in terms of selectivity.

4.1 THE SUBSPACE-BASED ADAPTIVE SIDELOBE BLANKER (S-ASB)

As stated in Section 3.4.2, the ASB is the cascade of the AMF and the ACE detectors; remarkably, such a two-stage detector has adjustable directivity. Obviously, better sidelobe signals rejection is gained at the price of some loss on mainlobe targets [17, 44, 46].

We present a modification of the ASB algorithm that increases its operational range in robustness, while retaining an acceptable detection loss with respect to Kelly's detector, which, as already stated, is considered as the benchmark detector in case of matched signals. To this end, we replace the robust stage of the ASB, i.e., the AMF, with the SD. Recall that the decision statistic of the SD is given by

$$t_{\mathrm{SD}} = \frac{\boldsymbol{r}^\dagger \boldsymbol{S}^{-1} \boldsymbol{H} (\boldsymbol{H}^\dagger \boldsymbol{S}^{-1} \boldsymbol{H})^{-1} \boldsymbol{H}^\dagger \boldsymbol{S}^{-1} \boldsymbol{r}}{1 + \boldsymbol{r}^\dagger \boldsymbol{S}^{-1} \boldsymbol{r}},$$

where $\boldsymbol{H} = [\boldsymbol{v}\ \boldsymbol{v}_1 \cdots \boldsymbol{v}_{r-1}] \in \mathbb{C}^{N\times r}$ is a full-column-rank matrix (and, hence, $r \geq 1$ is the rank of $\boldsymbol{H}$). Obviously, the choice of $\boldsymbol{H}$ will impact on the performance of the overall detector: for matched signals the lower the value of r the better the performance; however, values of r greater than one increase the robustness of the overall detector in presence of mainlobe targets. We will show that the choice $\boldsymbol{H} = [\boldsymbol{v}\ \boldsymbol{v}_1]$, with $\boldsymbol{v}_1$ a vector slightly mismatched with respect to $\boldsymbol{v}$, guarantees an enhanced robustness with respect to the ASB in homogeneous environment.

Summarizing, the operation of this two-stage detector, referred to in the following as Subspace-based ASB (S-ASB), can be pictorially described as in Figure 4.1 where t_{ACE} denotes

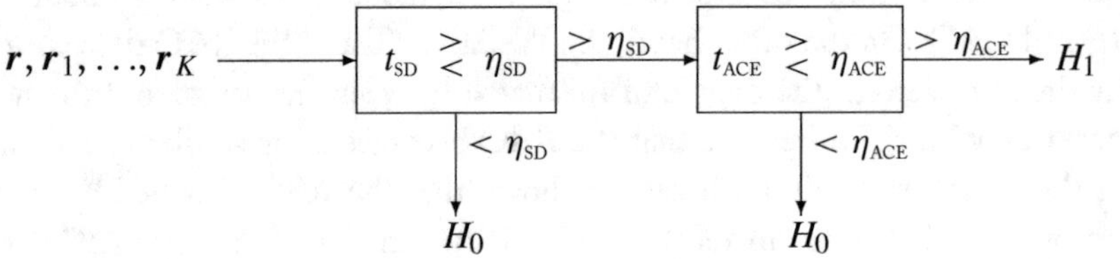

Figure 4.1: Block diagram of the S-ASB.

the decision statistic of the ACE, given by equation (2.22), and $(\eta_{\mathrm{SD}}, \eta_{\mathrm{ACE}})$ is the threshold pair to be set in order to guarantee a preassigned P_{fa}.

Note that $0 \leq t_{\mathrm{SD}} < 1$ and $0 \leq t_{\mathrm{ACE}} \leq 1$ by construction; as a consequence, meaningful detection thresholds satisfy the constraints $0 \leq \eta_{\mathrm{SD}} < 1$ and $0 \leq \eta_{\mathrm{ACE}} \leq 1$. The effectiveness of the proposed strategy for $r > 1$ will be proved by assessing its performance also in comparison to the ASB.

4.1.1 P_{fa} AND P_d OF THE S-ASB

Derivation of closed-form expressions for the P_{fa} and the P_d of the S-ASB for matched and mismatched signals can be easily accomplished in terms of the equivalent decision statistics

$$\tilde{t}_{\mathrm{SD}} = \frac{1}{1 - t_{\mathrm{SD}}} \quad \text{and} \quad \tilde{t}_{\mathrm{ACE}} = \frac{t_{\mathrm{ACE}}}{1 - t_{\mathrm{ACE}}}$$

of the robust and the selective stage, respectively. In fact, such statistics admit the following stochastic representations, derived in Appendices B and C,

$$\tilde{t}_{\mathrm{SD}} = (1 + c)\left(\tilde{t}_{\mathrm{K}} + 1\right), \tag{4.1}$$

$$\tilde{t}_{\mathrm{ACE}} = \tilde{t}_{\mathrm{K}} \left(1 + \frac{1}{c(1+b) + b}\right), \tag{4.2}$$

where

$$\tilde{t}_{\mathrm{K}} = \frac{t_{\mathrm{K}}}{1 - t_{\mathrm{K}}}, \tag{4.3}$$

with t_{K} given by (2.21).

Moreover, under the H_0 hypothesis (see also Appendix C),

- $\tilde{t}_{\mathrm{K}}$, given b and c, is ruled by a complex central F-distribution with $1, K-N+1$ degrees of freedom [2];
- b is a complex central F-distributed rv with $N-r, K-N+r+1$ degrees of freedom, i.e., $b \sim \mathcal{CF}_{N-r,K-N+r+1}$;
- $c \sim \mathcal{CF}_{r-1,K-N+2}$;
- b and c are statistically independent rv's.

Observe that the distributions of $\tilde{t}_{\mathrm{K}}$, b, and c do not depend on the true value of the covariance matrix $\boldsymbol{M}$; hence, the S-ASB possesses the CFAR property and its P_{fa} can be expressed as

$$P_{fa}(\eta_{\mathrm{SD}}, \eta_{\mathrm{ACE}}) = \mathrm{P}\left[t_{\mathrm{SD}} > \eta_{\mathrm{SD}}, t_{\mathrm{ACE}} > \eta_{\mathrm{ACE}};\ H_0\right]. \tag{4.4}$$

Inserting (4.1) and (4.2) into (4.4) yields

$$\begin{aligned}
P_{fa}(\eta_{\mathrm{SD}}, \eta_{\mathrm{ACE}}) &= \mathrm{P}[\tilde{t}_{\mathrm{SD}} > \tilde{\eta}_{\mathrm{SD}},\ \tilde{t}_{\mathrm{ACE}} > \tilde{\eta}_{\mathrm{ACE}};\ H_0] \\
&= \mathrm{P}\left[(1+c)\left(\tilde{t}_{\mathrm{K}}+1\right) > \tilde{\eta}_{\mathrm{SD}},\ \tilde{t}_{\mathrm{K}}\left(1+\frac{1}{c(1+b)+b}\right) > \tilde{\eta}_{\mathrm{ACE}};\ H_0\right] \\
&= \int_0^{+\infty}\int_0^{+\infty} \mathrm{P}\left[\tilde{t}_{\mathrm{K}} > \frac{\tilde{\eta}_{\mathrm{SD}}}{1+\gamma} - 1, \tilde{t}_{\mathrm{K}} > \tilde{\eta}_{\mathrm{ACE}}\frac{\gamma(1+\beta)+\beta}{(1+\beta)(1+\gamma)} | b=\beta, c=\gamma;\ H_0\right] \\
&\quad \times p_b(\beta)\, p_c(\gamma)\, d\beta\, d\gamma \\
&= \int_0^{+\infty}\int_0^{+\infty} \mathrm{P}\left[\tilde{t}_{\mathrm{K}} > \max\left(\frac{\tilde{\eta}_{\mathrm{SD}}}{1+\gamma} - 1, \tilde{\eta}_{\mathrm{ACE}}\frac{\gamma(1+\beta)+\beta}{(1+\beta)(1+\gamma)}\right) | b=\beta, c=\gamma;\ H_0\right] \\
&\quad \times p_b(\beta)\, p_c(\gamma)\, d\beta\, d\gamma \\
&= 1 - \int_0^{+\infty}\int_0^{+\infty} \mathcal{P}_0\left(\max\left(\frac{\tilde{\eta}_{\mathrm{SD}}}{1+\gamma} - 1, \tilde{\eta}_{\mathrm{ACE}}\frac{\gamma(1+\beta)+\beta}{(1+\beta)(1+\gamma)}\right)\right) \\
&\quad \times p_b(\beta)\, p_c(\gamma)\, d\beta\, d\gamma,
\end{aligned} \tag{4.5}$$

where

$$\tilde{\eta}_{\mathrm{SD}} = \frac{1}{1-\eta_{\mathrm{SD}}}, \quad \tilde{\eta}_{\mathrm{ACE}} = \frac{\eta_{\mathrm{ACE}}}{1-\eta_{\mathrm{ACE}}}, \tag{4.6}$$

$p_b(\cdot)$ is the pdf of the rv $b \sim \mathcal{CF}_{N-r,K-N+r+1}$, $p_c(\cdot)$ is the pdf of the rv $c \sim \mathcal{CF}_{r-1,K-N+2}$, and $\mathcal{P}_0(\cdot)$ is the CDF of the rv $\tilde{t}_{\mathrm{K}}$, given b and c (and under H_0), i.e., the CDF of a rv ruled by the $\mathcal{CF}_{1,K-N+1}$ distribution. Note also that

$$1 \leq \tilde{\eta}_{\mathrm{SD}} < +\infty \quad \text{and} \quad 0 \leq \tilde{\eta}_{\mathrm{ACE}} < +\infty.$$

A few remarks are now in order. First, P_{fa} depends on the two thresholds η_{SD} and η_{ACE}, as explicitly indicated in the left-most side of (4.5); more important, there exist infinite threshold pairs

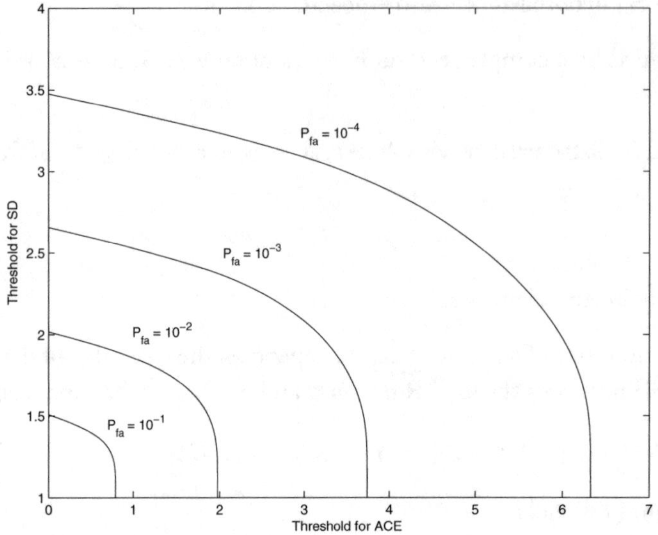

Figure 4.2: Contours of constant P_{fa} for the S-ASB, $N = 8$, $K = 16$, and $r = 2$.

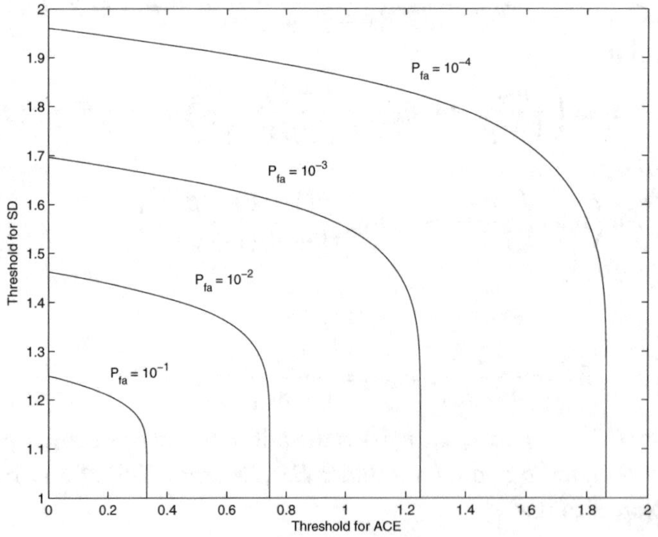

Figure 4.3: Contours of constant P_{fa} for the S-ASB, $N = 16$, $K = 32$, and $r = 2$.

η_{SD} and η_{ACE} or, equivalently, $\tilde{\eta}_{\mathrm{SD}}$ and $\tilde{\eta}_{\mathrm{ACE}}$, that guarantee a desired value of P_{fa}. Figures 4.2 and 4.3 show the contour plots corresponding to different values of P_{fa} in the plane $\tilde{\eta}_{\mathrm{ACE}}$ $\tilde{\eta}_{\mathrm{SD}}$. Second, it shows that the threshold setting necessary to guarantee a preassigned P_{fa} is independent of $\boldsymbol{H}$ but for the value of r. Finally, equation (4.5) encompasses the following special cases

$$P_{fa}(\eta_{\mathrm{SD}}, \eta_{\mathrm{ACE}}) = \begin{cases} P_{fa} \text{ of the ACE}, & \text{if } \tilde{\eta}_{\mathrm{SD}} = 1, \tilde{\eta}_{\mathrm{ACE}} > 0, \\ P_{fa} \text{ of the SD}, & \text{if } \tilde{\eta}_{\mathrm{SD}} > 1, \tilde{\eta}_{\mathrm{ACE}} = 0, \\ 1, & \text{if } \tilde{\eta}_{\mathrm{SD}} = 1, \tilde{\eta}_{\mathrm{ACE}} = 0. \end{cases}$$

Under the H_1 hypothesis, we assume a possible misalignment between the actual steering vector $\boldsymbol{p}$ and the nominal one $\boldsymbol{v}$ (see equation (3.1)). In this case, the rv's b and c depend on the mismatch angle θ, defined through equation (3.6), between the actual steering vector and the nominal one in the whitened observation space; for this reason, in the following, we will denote these rv's by b_θ and c_θ. Due to the presence of signal components, the distributions of $\tilde{t}_{\mathrm{K}}$, b_θ, and c_θ change, more precisely (see Appendix C)

- $\tilde{t}_{\mathrm{K}}$, given b_θ and c_θ, is ruled by the complex noncentral F-distribution with $1, K-N+1$ degrees of freedom and noncentrality parameter δ_θ, with

  $$\delta_\theta^2 = \frac{\mathrm{SNR}\cos^2\theta}{(1+b_\theta)(1+c_\theta)},$$

  where we recall that

  $$\mathrm{SNR} = |\alpha|^2 \boldsymbol{p}^\dagger \boldsymbol{M}^{-1} \boldsymbol{p}$$

  is the total available signal-to-noise ratio;

- b_θ is ruled by the complex noncentral F-distribution with $N-r, K-N+r+1$ degrees of freedom and noncentrality parameter δ_{b_θ}, with

  $$\delta_{b_\theta}^2 = \mathrm{SNR}\ \sin^2\theta\ \|\boldsymbol{h}_C\|^2,$$

  i.e., $b_\theta \sim \mathcal{CF}_{N-r,K-N+r+1}(\delta_{b_\theta})$, where $\boldsymbol{h}_C$ has been defined in Chapter 3;

- given b_θ, $c_\theta \sim \mathcal{CF}_{r-1,K-N+2}(\delta_{c_\theta})$, with

  $$\delta_{c_\theta}^2 = \frac{\mathrm{SNR}\ \sin^2\theta\ \|\boldsymbol{h}_B\|^2}{1+b_\theta},$$

  where $\boldsymbol{h}_B$ has been defined in Chapter 3.

Thus, proceeding along the same line as for the derivation of the P_{fa}, it is easy to see that the P_d is given by

$$
\begin{aligned}
&P_d(\mathrm{SNR}, \eta_{\mathrm{SD}}, \eta_{\mathrm{ACE}}, \cos^2\theta, \|\boldsymbol{h}_B\|^2) = \mathrm{P}\left[t_{\mathrm{SD}} > \eta_{\mathrm{SD}}, t_{\mathrm{ACE}} > \eta_{\mathrm{ACE}};\ H_1\right] \\
&= \mathrm{P}\left[\tilde{t}_{\mathrm{SD}} > \tilde{\eta}_{\mathrm{SD}}, \tilde{t}_{\mathrm{ACE}} > \tilde{\eta}_{\mathrm{ACE}};\ H_1\right] \\
&= \mathrm{P}\left[(1+c_\theta)(\tilde{t}_{\mathrm{K}}+1) > \tilde{\eta}_{\mathrm{SD}},\ \tilde{t}_{\mathrm{K}}\left(1+\frac{1}{c_\theta(1+b_\theta)+b_\theta}\right) > \tilde{\eta}_{\mathrm{ACE}};\ H_1\right] \\
&= \int_0^{+\infty}\int_0^{+\infty} \mathrm{P}\left[\tilde{t}_{\mathrm{K}} > \frac{\tilde{\eta}_{\mathrm{SD}}}{1+\gamma} - 1, \tilde{t}_{\mathrm{K}} > \tilde{\eta}_{\mathrm{ACE}}\frac{\gamma(1+\beta)+\beta}{(1+\beta)(1+\gamma)} | b_\theta = \beta, c_\theta = \gamma;\ H_1\right] \\
&\times p_{b_\theta c_\theta}(\beta,\gamma)\, d\beta\, d\gamma \\
&= 1 - \int_0^{+\infty}\int_0^{+\infty} \mathcal{P}_1\left(\max\left(\frac{\tilde{\eta}_{\mathrm{SD}}}{1+\gamma} - 1, \tilde{\eta}_{\mathrm{ACE}}\frac{\gamma(1+\beta)+\beta}{(1+\beta)(1+\gamma)}\right)\right) \\
&\times p_{c_\theta|b_\theta}(\gamma|b_\theta=\beta)\, p_{b_\theta}(\beta)\, d\beta\, d\gamma,
\end{aligned}
\tag{4.7}
$$

where $\mathcal{P}_1(\cdot)$ is the CDF of the rv $\tilde{t}_{\mathrm{K}}$, given b_θ and c_θ (and under H_1), i.e., the CDF of a rv ruled by the $\mathcal{CF}_{1,K-N+1}(\delta_\theta)$ distribution, $p_{b_\theta}(\cdot)$ is the pdf of a rv ruled by the $\mathcal{CF}_{N-r,K-N+r+1}(\delta_{b_\theta})$, and $p_{c_\theta|b_\theta}(\cdot|\cdot)$ is the pdf of a rv ruled by the $\mathcal{CF}_{r-1,K-N+2}(\delta_{c_\theta})$.

In the case of perfect match between $\boldsymbol{v}$ and $\boldsymbol{p}$, i.e., $\theta = 0$, δ_{b_θ} and δ_{c_θ} are equal to zero; thus the rv's c_θ and b_θ obey the complex central F-distribution with $N-r$, $K-N+r+1$ and $r-1$, $K-N+2$ degrees of freedom, respectively. On the other hand, $\tilde{t}_{\mathrm{K}}$ is still subject to the complex noncentral F-distribution with 1, $K-N+1$ degrees of freedom and noncentrality parameter given by

$$
\delta_0^2 = \frac{\mathrm{SNR}}{(1+b)(1+c)}.
$$

Note that the dependence of P_d on the signal parameters is entirely confined to the defined signal-to-noise ratio SNR, $\cos^2\theta$, $\|\boldsymbol{h}_B\|^2$ (or $\|\boldsymbol{h}_C\|^2$). Finally, equation (4.7) encompasses the following limiting expressions

$$
P_d = \begin{cases} P_d \text{ of the ACE}, & \text{if } \tilde{\eta}_{\mathrm{SD}} = 1, \tilde{\eta}_{\mathrm{ACE}} > 0, \\ P_d \text{ of the SD}, & \text{if } \tilde{\eta}_{\mathrm{SD}} > 1, \tilde{\eta}_{\mathrm{ACE}} = 0, \\ 1, & \text{if } \tilde{\eta}_{\mathrm{SD}} = 1, \tilde{\eta}_{\mathrm{ACE}} = 0. \end{cases}
$$

Formula (4.7) also highlights that P_d for matched signals is independent of $\boldsymbol{H}$, but for the value of r (provided that the space spanned by the columns of $\boldsymbol{H}$ contains the nominal steering vector $\boldsymbol{v}$).

Figures 4.4 and 4.5 show detection performance of the S-ASB in case of matched signals with $N = 16$, $K = 32$, $P_{fa} = 10^{-4}$, and different values of r. More precisely, in Figure 4.4 the P_d is evaluated for those threshold pairs which yield approximately the minimum loss with respect to Kelly's detector for $P_d = 0.9$, $P_{fa} = 10^{-4}$ (see Table 4.1 for the threshold values). On the other hand, in Figure 4.5 the threshold pairs are chosen in order to achieve approximately the maximum loss with respect to Kelly's detector for $P_d = 0.9$, $P_{fa} = 10^{-4}$ (see Table 4.1 for the threshold values). Inspection of these figures confirms that, in case of perfectly aligned signals, the performance of the

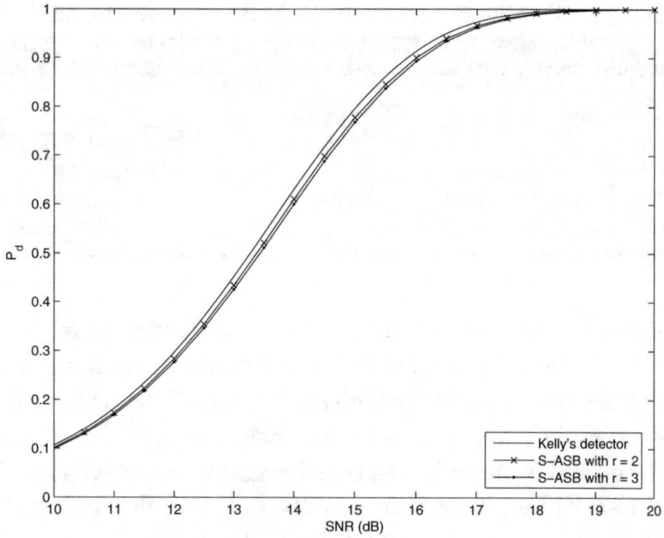

Figure 4.4: P_d versus SNR for the S-ASB and Kelly's detector with $N = 16$, $K = 32$, $P_{fa} = 10^{-4}$, and threshold pairs corresponding to the minimum loss of the S-ASB with respect to Kelly's detector.

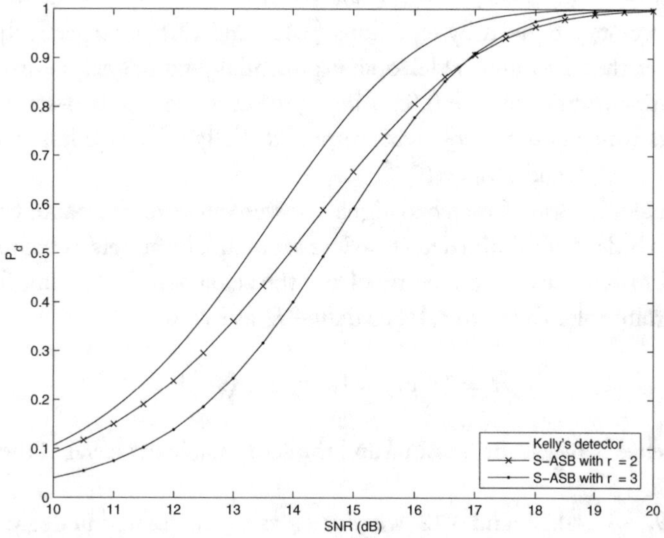

Figure 4.5: P_d versus SNR for the S-ASB and Kelly's detector with $N = 16$, $K = 32$, $P_{fa} = 10^{-4}$, and threshold pairs corresponding to the maximum loss of the S-ASB with respect to Kelly's detector.

Table 4.1: Threshold pairs of the S-ASB in Figures 4.4 and 4.5.

r	Minimum Loss (Fig. 4.4)	Maximum Loss (Fig. 4.5)
$r = 2$	$\tilde{\eta}_{\mathrm{SD}} = 1.81, \tilde{\eta}_{\mathrm{ACE}} = 1.30$	$\tilde{\eta}_{\mathrm{SD}} = 1, \tilde{\eta}_{\mathrm{ACE}} = 1.86$
$r = 3$	$\tilde{\eta}_{\mathrm{SD}} = 1.89, \tilde{\eta}_{\mathrm{ACE}} = 1.36$	$\tilde{\eta}_{\mathrm{SD}} = 2.18, \tilde{\eta}_{\mathrm{ACE}} = 0$

S-ASB depends on r and shows that the lower the value of r the better the performance. Figures 4.6 and 4.7 plot P_d vs SNR for the ASB and the S-ASB, respectively, assuming $N = 8$, $K = 16$, and $r = 2$. In these figures there are two curves for both the S-ASB and the ASB: such curves correspond to the limiting behaviors of the two detectors for threshold settings which guarantee $P_{fa} = 10^{-4}$. From inspection of the quoted figures it is seen that the maximum loss (of about 2 dB) at $P_d = 0.9$ of the S-ASB with respect to Kelly's detector is similar to that of the ASB; such a maximum loss decreases to (about) 1 dB in Figures 4.5 and 4.8 for both detectors.

4.1.2 PERFORMANCE PREDICTION IN PRESENCE OF MISMATCHED SIGNALS

In this section we investigate the performance of the S-ASB in presence of mismatched signals. To this end, assume $N_p = 1$ and, hence, $N = N_a$ and recall that the expressions for the nominal and the actual steering vectors are given by equations (3.11) and (3.12), respectively. As to the noise vectors, we remember that they are modeled as exponentially-correlated complex normal vectors with one-lag correlation coefficient $\rho = 0.95$. Finally, the curves of the two-stage detectors refer to those threshold pairs which ensure a loss with respect to Kelly's detector less than about 1 dB for matched signals at $P_d = 0.9$ and $P_{fa} = 10^{-4}$.

As already stated, in case of matched signals, lower values of the parameter r return better performance; thus, in order to limit the detection loss for mainlobe targets, it seems reasonable to set $r = 2$. In case of mismatch between the nominal and the actual steering vector, the second column of $\boldsymbol{H}$ plays an important role. To see this, let us define $\boldsymbol{H}$ as follows

$$\boldsymbol{H} = [\boldsymbol{v}\ \boldsymbol{v}_1] = [\boldsymbol{a}(\pi/2)\ \boldsymbol{a}(\psi_1)]$$

and ψ_T as the difference between the nominal and the actual angle of arrival of the impinging signal echo.

In Figures 4.9, 4.10, 4.11, and 4.12 we plot P_d vs ψ_T (measured in degrees) for the S-ASB with $r = 2$ and four choices of $\psi_1 \in (0, \pi/2)$ (actually values of ψ_1 not too far from $\pi/2$), assuming $N = 16$, $K = 32$, $P_{fa} = 10^{-4}$, and SNR $= 19$ dB. The figures highlight that $\psi_1 \leq \pi/2 - 4\pi/360$ produce a somewhat undesirable behavior of some curves whose transition from the "bandpass" to the "stopband" is not sharp. From this point of view values of $\psi_1 \geq \pi/2 - 2\pi/360$ are to be

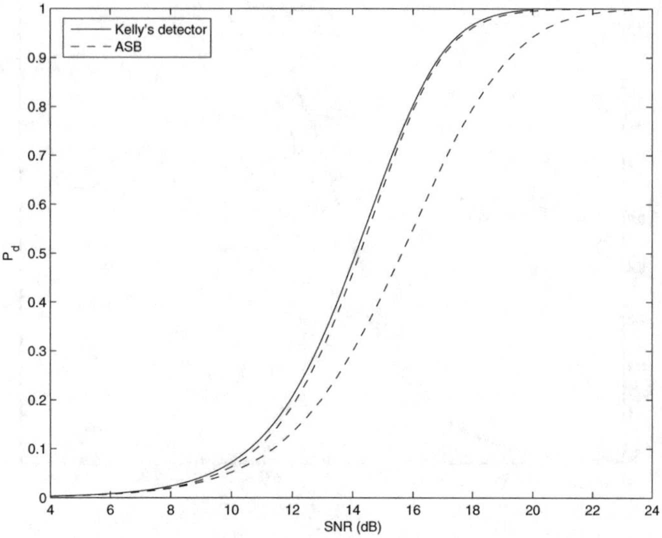

Figure 4.6: P_d versus SNR for the ASB and Kelly's detector with $N = 8$, $K = 16$, and $P_{fa} = 10^{-4}$.

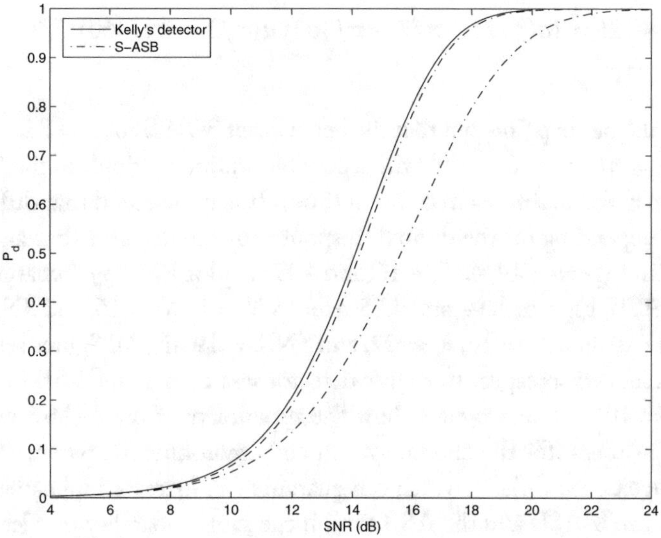

Figure 4.7: P_d versus SNR for the S-ASB and Kelly's detector with $N = 8$, $K = 16$, $r = 2$, and $P_{fa} = 10^{-4}$.

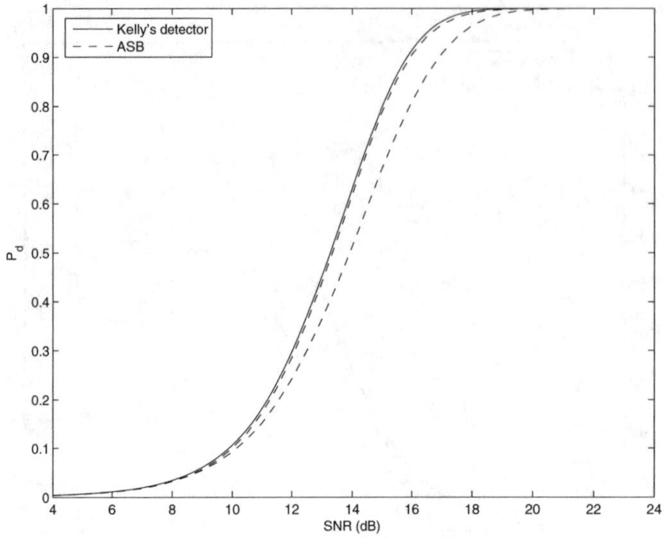

Figure 4.8: P_d versus SNR for the ASB and Kelly's detector with $N = 16$, $K = 32$, and $P_{fa} = 10^{-4}$.

preferred. Analogous curves for $r = 3$ are shown in Figure 4.13, where

$$\boldsymbol{H} = [\boldsymbol{a}(\pi/2)\ \boldsymbol{a}(\pi/2 - \pi/360)\ \boldsymbol{a}(\pi/2 - 2\pi/360)].$$

Inspection of the figure points out that the behavior of S-ASB for $r = 3$ is somehow reminiscent of $r = 2$ and $\boldsymbol{v}_1 = \boldsymbol{a}(\pi/2 - 6\pi/360)$ with a possible additional ripple in the "bandpass" region. Hereafter we assume $r = 2$ and $\boldsymbol{v}_1 = \boldsymbol{a}(\pi/2 - \pi/360)$. It is understood that different values of $\boldsymbol{v}_1$ could be of interest depending on the desired "response" of the detector (but also on the adopted system parameters). In Figures 4.14, 4.15, 4.16, and 4.17 we plot P_d vs ψ_T (measured in degrees) for the S-ASB and the ASB; Figures 4.14 and 4.15 refer to $N = 8$, $K = 16$, and SNR $= 20$ dB, while Figures 4.16 and 4.17 assume $N = 16$, $K = 32$, and SNR $= 19$ dB. All figures set the thresholds in order to guarantee a loss with respect to Kelly's detector less than about 1 dB for matched signals, $P_d = 0.9$, and $P_{fa} = 10^{-4}$. These figures show the superiority of the S-ASB with respect to the ASB in terms of robustness (for the chosen system and environmental parameters). This result is due to the robust stage of the S-ASB, which can guarantee an increased robustness with respect to the AMF. However, the S-ASB and the ASB exhibit the same capability to reject sidelobe signals, according to the fact that the selective stage (the ACE) is the same. In next section, we present a modified version of the S-ASB obtained replacing the ACE with a more selective receiver in order to increase the sidelobe rejection capabilities of the two-stage detector.

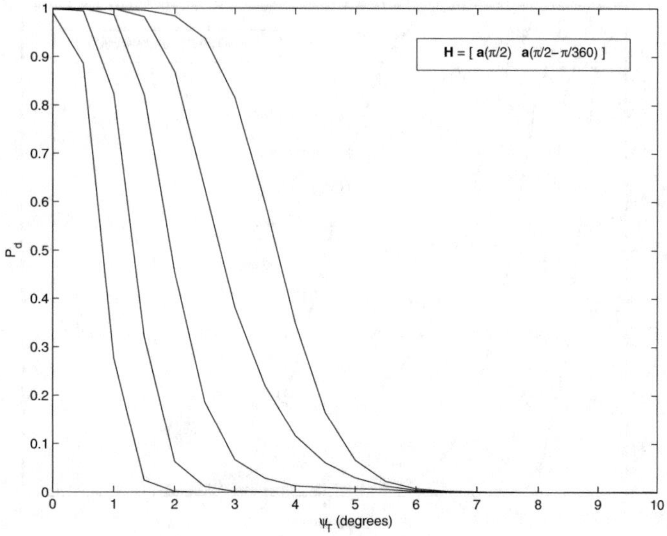

Figure 4.9: P_d versus ψ_T for the S-ASB with $N = 16$, $K = 32$, $P_{fa} = 10^{-4}$, SNR= 19 dB, and $\psi_1 = \pi/2 - \pi/360$.

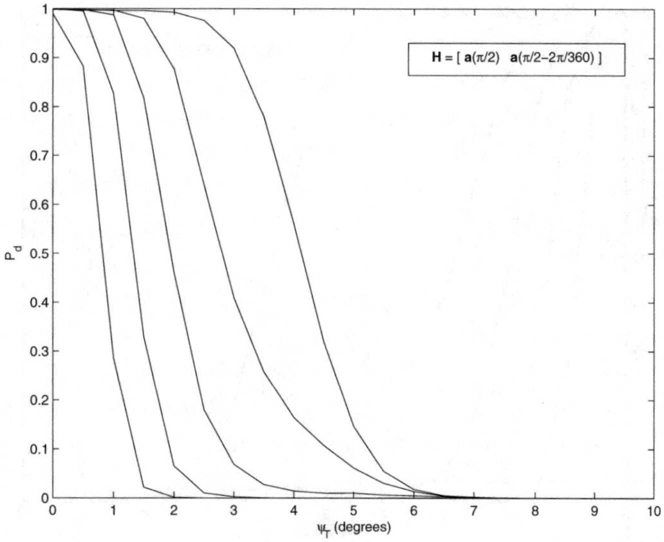

Figure 4.10: P_d versus ψ_T for the S-ASB with $N = 16$, $K = 32$, $P_{fa} = 10^{-4}$, SNR= 19 dB, and $\psi_1 = \pi/2 - 2\pi/360$.

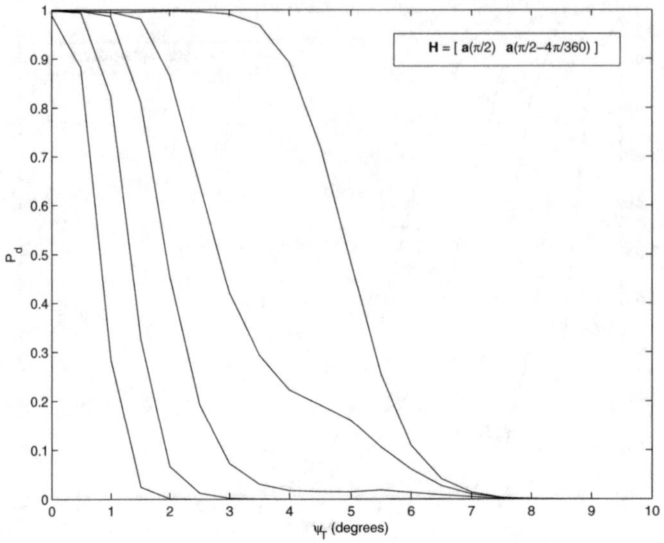

Figure 4.11: P_d versus ψ_T for the S-ASB with $N = 16$, $K = 32$, $P_{fa} = 10^{-4}$, SNR= 19 dB, and $\psi_1 = \pi/2 - 4\pi/360$.

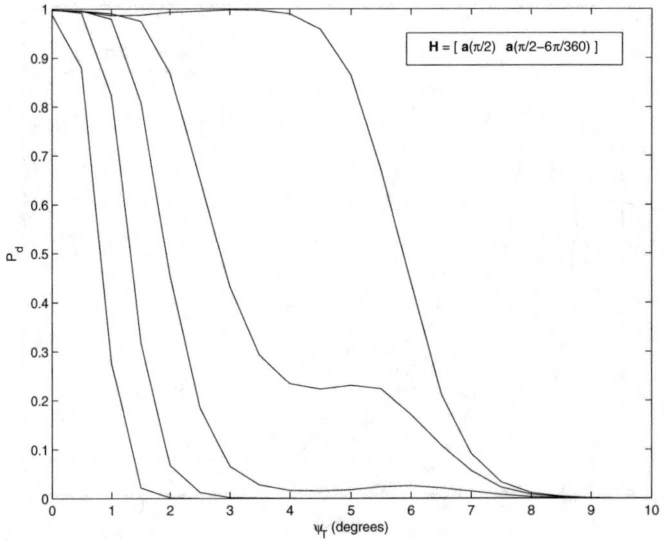

Figure 4.12: P_d versus ψ_T for the S-ASB with $N = 16$, $K = 32$, $P_{fa} = 10^{-4}$, SNR= 19 dB, and $\psi_1 = \pi/2 - 6\pi/360$.

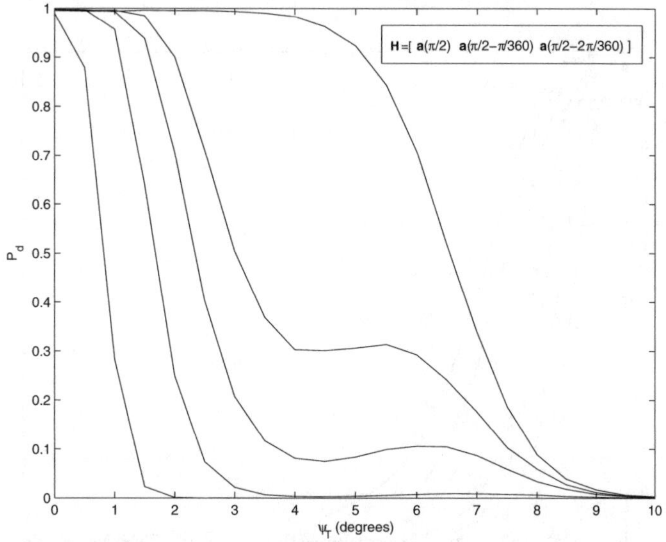

Figure 4.13: P_d versus ψ_T for the S-ASB with $N = 16$, $K = 32$, $P_{fa} = 10^{-4}$, and SNR= 19 dB.

4.2 THE W-ABORT AND SUBSPACE-BASED ADAPTIVE SIDELOBE BLANKER (WAS-ASB)

The comparisons among selective receivers made in Chapter 3 pointed out that the W-ABORT ensures the best rejection capabilities of unwanted signals, while having a moderate detection loss with respect to Kelly's detector in case of matched signals. As a matter of fact, it is possible to extend the operational range of the S-ASB in selectivity by replacing its selective stage with the W-ABORT, whose statistic is given by (3.16). A block scheme of this detector, referred to in the following as W-ABORT and Subspace-based Adaptive Sidelobe Blanker (WAS-ASB), is drawn in Figure 4.18, where t_{WA} denotes the decision statistic of the W-ABORT and $(\eta_{\text{SD}}, \eta_{\text{WA}})$ is the threshold pair to be set in order to guarantee a preassigned P_{fa}.

As for S-ASB, we assume that the matrix $\boldsymbol{H} \in \mathbb{C}^{N\times r}$, defined as

$$\boldsymbol{H} = [\boldsymbol{v}\ \boldsymbol{v}_1\ \ldots\ \boldsymbol{v}_{r-1}],$$

is a full-column-rank matrix. It is obvious that the performance of the overall detector will depend on the structure of $\boldsymbol{H}$ (see also Section 4.1).

4.2.1 P_{fa} AND P_d OF THE WAS-ASB

In this section we derive analytical expressions for P_d and P_{fa} of the WAS-ASB. To this end, we replace t_{SD} with the equivalent decision statistic $\tilde{t}_{\text{SD}}$ defined at the beginning of Section 4.1.1. Moreover, based upon results contained in Appendices B and C, it is possible to show that t_{WA} admits

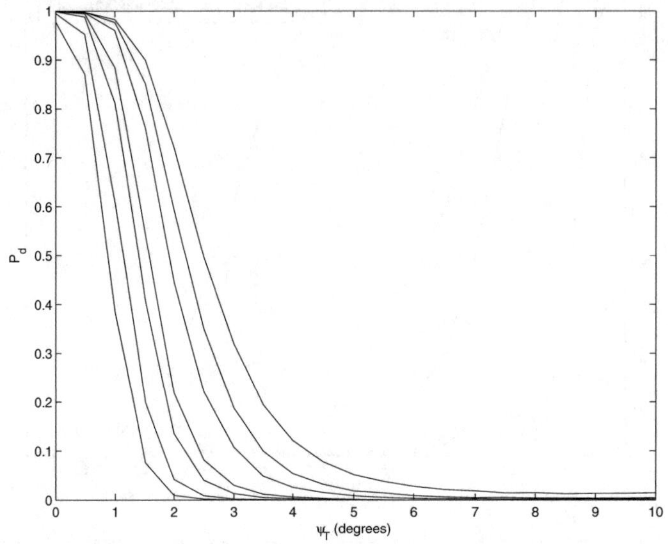

Figure 4.14: P_d versus ψ_T for the ASB with $N = 8$, $K = 16$, $P_{fa} = 10^{-4}$, and SNR= 20 dB.

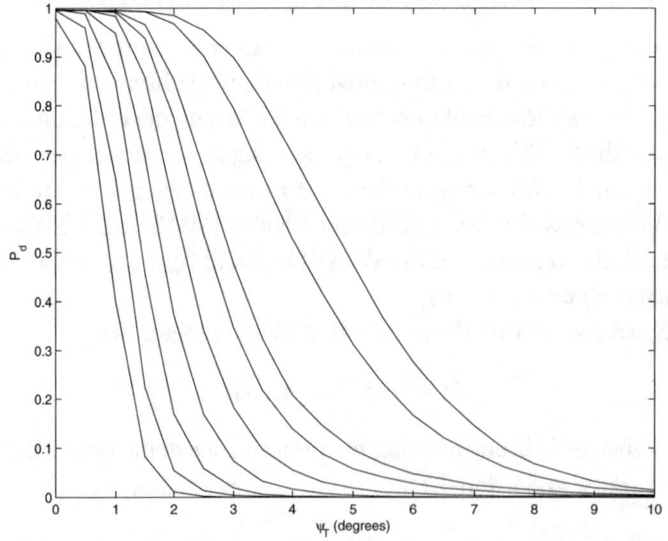

Figure 4.15: P_d versus ψ_T for the S-ASB with $N = 8$, $K = 16, r = 2$, $P_{fa} = 10^{-4}$, and SNR= 20 dB.

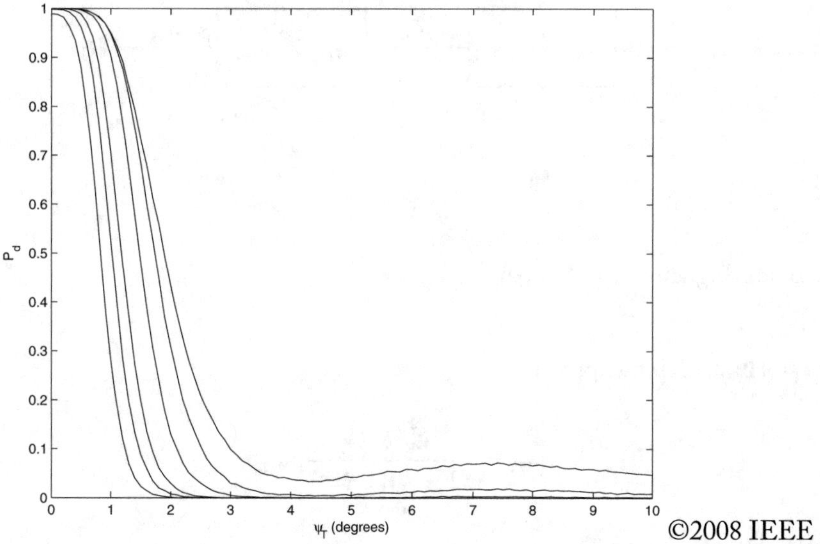

Figure 4.16: P_d versus ψ_T for the ASB with $N = 16$, $K = 32$, $P_{fa} = 10^{-4}$, and SNR= 19 dB.

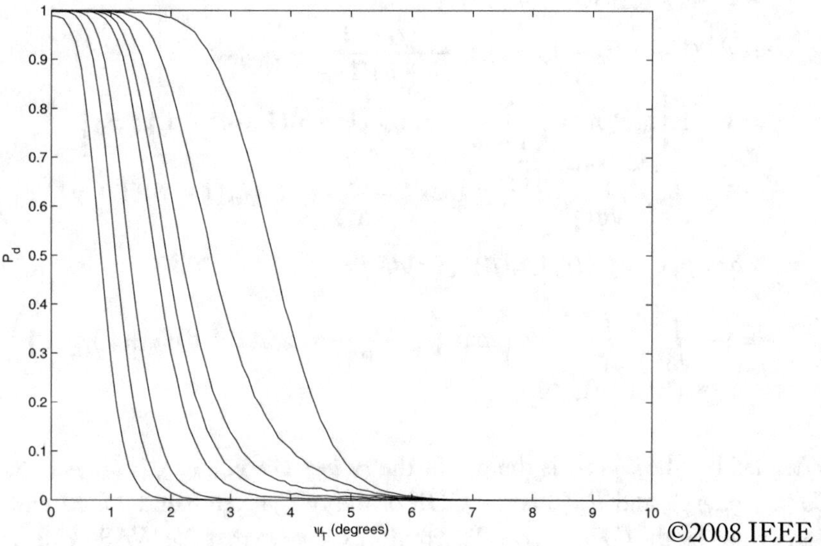

Figure 4.17: P_d versus ψ_T for the S-ASB with $N = 16$, $K = 32$, $r = 2$, $P_{fa} = 10^{-4}$, and SNR= 19 dB.

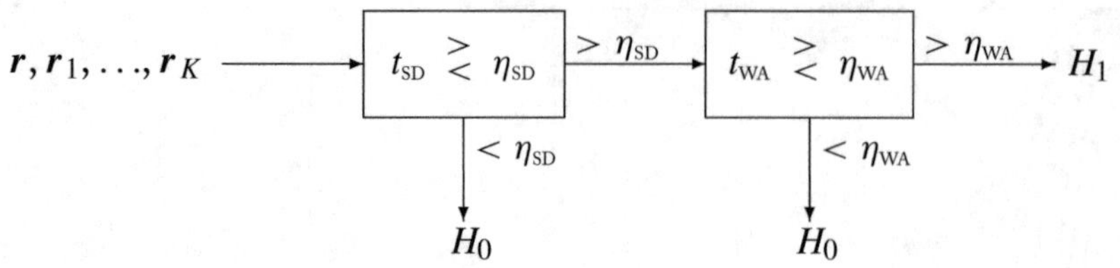

Figure 4.18: Block diagram of the WAS-ASB.

the following stochastic representation

$$t_{\mathrm{WA}} = \frac{(\tilde{t}_{\mathrm{K}} + 1)}{(1+b)(1+c)},$$

where $\tilde{t}_{\mathrm{K}}$ is given by equation (4.3).

Thus, it is straightforward to see that the P_{fa} is given by

$$\begin{aligned}
P_{fa}(\eta_{\mathrm{SD}}, \eta_{\mathrm{WA}}) &= \mathrm{P}\left[t_{\mathrm{SD}} > \eta_{\mathrm{SD}}, t_{\mathrm{WA}} > \eta_{\mathrm{WA}}; H_0\right] \\
&= \mathrm{P}\left[\tilde{t}_{\mathrm{SD}} > \tilde{\eta}_{\mathrm{SD}}, t_{\mathrm{WA}} > \eta_{\mathrm{WA}}; H_0\right] \\
&= \mathrm{P}\left[(1+c)(\tilde{t}_{\mathrm{K}}+1) > \tilde{\eta}_{\mathrm{SD}}, \frac{\tilde{t}_{\mathrm{K}}+1}{(1+b)(1+c)} > \eta_{\mathrm{WA}}; H_0\right] \\
&= 1 - \mathrm{P}\left[\tilde{t}_{\mathrm{K}} \le \max\left(\frac{\tilde{\eta}_{\mathrm{SD}}}{1+c} - 1, \eta_{\mathrm{WA}}(1+b)(1+c) - 1\right); H_0\right] \\
&= 1 - \int_0^{+\infty}\int_0^{+\infty} \mathrm{P}\left[\tilde{t}_{\mathrm{K}} \le \max\left(\frac{\tilde{\eta}_{\mathrm{SD}}}{1+\gamma} - 1, \eta_{\mathrm{WA}}(1+\beta)(1+\gamma) - 1\right) \mid \right. \\
&\qquad \left. b=\beta, c=\gamma; H_0\right] p_b(\beta)p_c(\gamma)d\beta d\gamma \\
&= 1 - \int_0^{+\infty}\int_0^{+\infty} \mathcal{P}_0\left(\max\left(\frac{\tilde{\eta}_{\mathrm{SD}}}{1+\gamma} - 1, \eta_{\mathrm{WA}}(1+\beta)(1+\gamma) - 1\right)\right) \\
&\qquad \times p_b(\beta)p_c(\gamma)d\beta d\gamma,
\end{aligned} \tag{4.8}$$

where $\tilde{\eta}_{\mathrm{SD}}$ is defined by (4.6), $p_b(\cdot)$ is the pdf of the rv $b \sim \mathcal{CF}_{N-r,K-N+r+1}$, $p_c(\cdot)$ is the pdf of the rv $c \sim \mathcal{CF}_{r-1,K-N+2}$, and $\mathcal{P}_0(\cdot)$ is the CDF of the rv $\tilde{t}_{\mathrm{K}}$, given b and c (and under H_0), i.e., the CDF of a rv ruled by the $\mathcal{CF}_{1,K-N+1}$ distribution. Observe that the WAS-ASB possesses the CFAR property with respect to $\boldsymbol{M}$; in fact, the above expression of P_{fa} can be computed without knowledge of the noise covariance matrix $\boldsymbol{M}$. For the reader's ease Figure 4.19 shows the contour plots for the WAS-ASB corresponding to different values of P_{fa}, as functions of the threshold pairs $(\eta_{\mathrm{WA}}, \tilde{\eta}_{\mathrm{SD}})$ with $N = 16$, $K = 32$, and $r = 2$. Finally, note that equation (4.8) encompasses the following special cases

$$P_{fa}(\eta_{\text{SD}}, \eta_{\text{WA}}) = \begin{cases} P_{fa} \text{ of the W-ABORT}, & \text{if } \tilde{\eta}_{\text{SD}} = 1, \eta_{\text{WA}} > 0, \\ P_{fa} \text{ of the SD}, & \text{if } \tilde{\eta}_{\text{SD}} > 1, \eta_{\text{WA}} = 0, \\ 1, & \text{if } \tilde{\eta}_{\text{SD}} = 1, \eta_{\text{WA}} = 0. \end{cases}$$

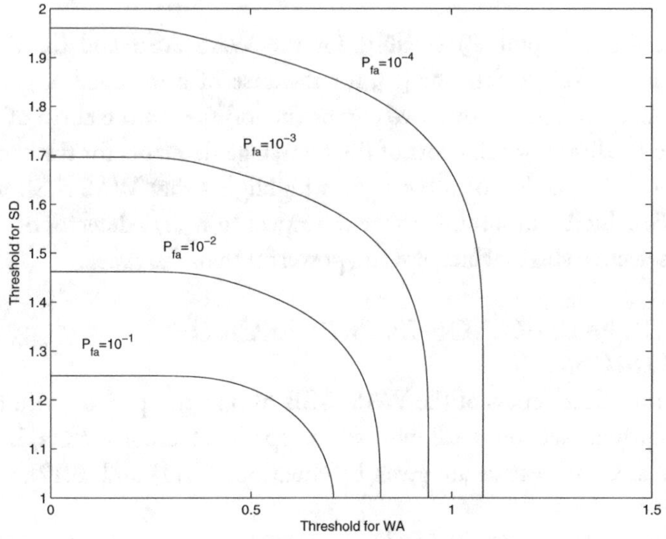

Figure 4.19: Contours of constant P_{fa} for the WAS-ASB with $N = 16$, $K = 32$, and $r = 2$.

As to the P_d, it is easy to see that for the general case of mismatched signals it is given by

$$\begin{aligned} P_d(\text{SNR}, \eta_{\text{SD}}, \eta_{\text{WA}}, \cos^2\theta, \|\boldsymbol{h}_B\|^2) &= \text{P}\left[t_{\text{SD}} > \eta_{\text{SD}}, t_{\text{WA}} > \eta_{\text{WA}};\ H_1\right] \\ &= \text{P}\left[\tilde{t}_{\text{SD}} > \tilde{\eta}_{\text{SD}},\ t_{\text{WA}} > \eta_{\text{WA}};\ H_1\right] \\ &= \text{P}\left[(1+c_\theta)(\tilde{t}_{\text{K}}+1) > \tilde{\eta}_{\text{SD}},\ \frac{\tilde{t}_{\text{K}}+1}{(1+b_\theta)(1+c_\theta)} > \eta_{\text{WA}};\ H_1\right] \\ &= 1 - \text{P}\left[\tilde{t}_{\text{K}} \le \max\left(\frac{\tilde{\eta}_{\text{SD}}}{1+c_\theta} - 1,\ \eta_{\text{WA}}(1+b_\theta)(1+c_\theta) - 1\right);\ H_1\right] \\ &= 1 - \int_0^{+\infty}\int_0^{+\infty} \mathcal{P}_1\left(\max\left(\frac{\tilde{\eta}_{\text{SD}}}{1+\gamma} - 1,\ \eta_{\text{WA}}(1+\beta)(1+\gamma) - 1\right)\right) \\ &\quad \times p_{b_\theta c_\theta}(\beta, \gamma)\, d\beta\, d\gamma \\ &= 1 - \int_0^{+\infty}\int_0^{+\infty} \mathcal{P}_1\left(\max\left(\frac{\tilde{\eta}_{\text{SD}}}{1+\gamma} - 1,\ \eta_{\text{WA}}(1+\beta)(1+\gamma) - 1\right)\right) \\ &\quad \times p_{c_\theta | b_\theta}(\gamma | b_\theta = \beta)\, p_{b_\theta}(\beta)\, d\beta\, d\gamma, \end{aligned} \tag{4.9}$$

where $\mathcal{P}_1(\cdot)$ is the CDF of the rv $\tilde{t}_{\text{K}}$, given b_θ and c_θ (and under H_1), i.e., the CDF of a rv ruled by the $\mathcal{CF}_{1,K-N+1}(\delta_\theta)$ distribution, $p_{b_\theta}(\cdot)$ is the pdf of a rv ruled by the $\mathcal{CF}_{N-r,K-N+r+1}(\delta_{b_\theta})$, and $p_{c_\theta|b_\theta}(\cdot|\cdot)$ is the pdf of a rv ruled by the $\mathcal{CF}_{r-1,K-N+2}(\delta_{c_\theta})$. Equation (4.9) encompasses the following special cases

$$P_d = \begin{cases} P_d \text{ of the W-ABORT}, & \text{if } \tilde{\eta}_{\mathrm{SD}} = 1,\ \eta_{\mathrm{WA}} > 0, \\ P_d \text{ of the SD}, & \text{if } \tilde{\eta}_{\mathrm{SD}} > 1,\ \eta_{\mathrm{WA}} = 0, \\ 1, & \text{if } \tilde{\eta}_{\mathrm{SD}} = 1,\ \eta_{\mathrm{WA}} = 0. \end{cases}$$

In Figures 4.20 and 4.21 we plot P_d vs SNR for the WAS-ASB and the AMF-RAO, respectively, as they compare to Kelly's detector [2], for the case of a matched target, assuming $N = 16$, $K = 32$, and $r = 2$; each figure shows two curves (in addition to the curve of Kelly's detector) which correspond to the limiting behaviors of the two-stage detectors for threshold settings which guarantee $P_{fa} = 10^{-4}$. Inspection of these figures highlights that WAS-ASB and AMF-RAO, in contrast to S-ASB, exhibit a maximum loss with respect to Kelly's detector of about 2 dB. Such a loss is due to their selective stages which are less powerful than the ACE.

4.2.2 PERFORMANCE PREDICTION IN PRESENCE OF MISMATCHED SIGNALS

In the following we assess the effectiveness of the WAS-ASB, also in comparison with the S-ASB and the AMF-RAO. Similarly to section 4.1.2, we assume spatial steering vectors, i.e., $N_p = 1$. Hence, nominal and actual steering vectors are given by equations (3.11) and (3.12), respectively. Moreover, we set

$$\boldsymbol{H} = [\boldsymbol{a}(\pi/2)\ \boldsymbol{a}(\pi/2 - \pi/360)].$$

Again, the noise vectors are modeled as exponentially-correlated complex normal ones with one-lag correlation coefficient $\rho = 0.95$.

In Figures 4.22 and 4.23, we plot P_d vs ψ_T (measured in degrees) for the WAS-ASB and the AMF-RAO, respectively, assuming $N = 16$, $K = 32$, and $P_{fa} = 10^{-4}$; for both detectors, plotted curves refer to the threshold pairs which ensure a loss with respect to Kelly's detector (and for matched signals) less than about 1 dB at $P_d = 0.9$. Observe from Figures 4.17 and 4.22 that the WAS-ASB can guarantee the same robustness of the S-ASB, due to the fact that they share the same robust stage, but better rejection capabilities than the S-ASB (and, consequently, better rejection capabilities than the ASB), due to the fact that the selective stage of the latter has been replaced by the W-ABORT. This result is more evident from Figure 4.24, where we plot contours of constant P_d for the threshold pairs corresponding to the most selective cases of Figures 4.17 and 4.22. Finally, by comparing Figures 4.22, 4.23, and 4.25, we see that the AMF-RAO is slightly more selective than the WAS-ASB for the considered parameter values.

However, the WAS-ASB can outperform the AMF-RAO detector in terms of selectivity, as confirmed by Figures 4.26 and 4.27, where we plot contours of constant P_d, as functions of SNR and $\cos^2\theta$, choosing thresholds which ensure a loss with respect to Kelly's detector less than about 1 dB for the perfectly matched case. More precisely, we compare the WAS-ASB with the AMF-RAO for $N = 16$, $K = 48$ in Figure 4.26 and for $N = 30$, $K = 60$ in Figure 4.27. The comparison shows

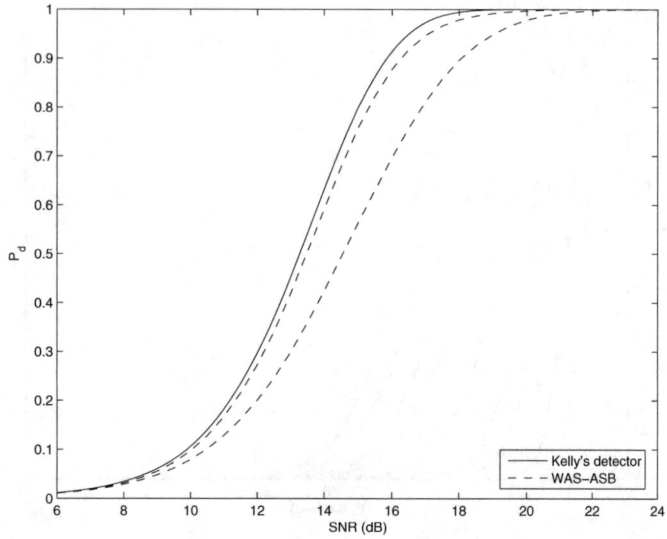

Figure 4.20: P_d versus SNR for the WAS-ASB with $N = 16$, $K = 32$, $r = 2$, and $P_{fa} = 10^{-4}$.

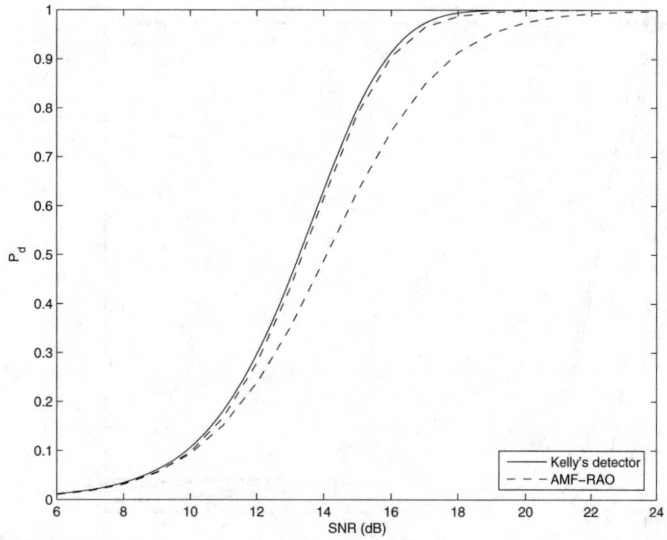

Figure 4.21: P_d versus SNR for the AMF-RAO with $N = 16$, $K = 32$, and $P_{fa} = 10^{-4}$.

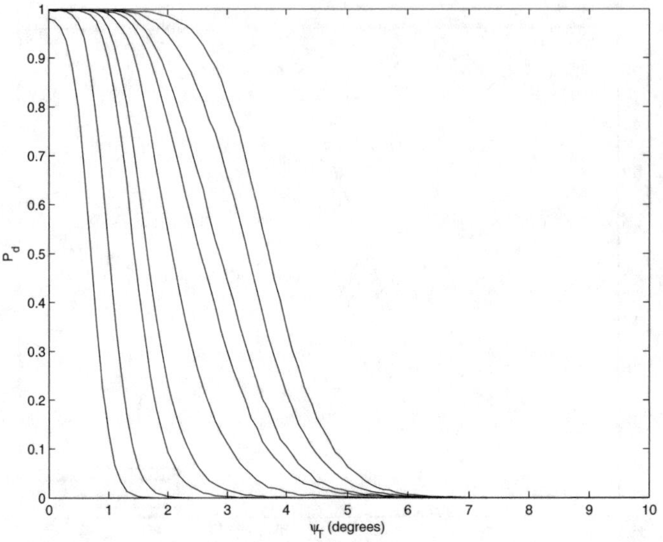

Figure 4.22: P_d versus ψ_T for the WAS-ASB with $N = 16, K = 32, r = 2, P_{fa} = 10^{-4}$, and SNR= 19 dB.

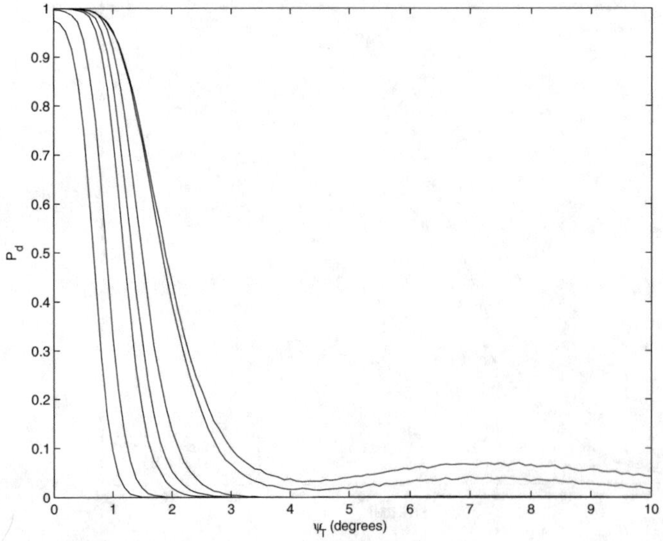

Figure 4.23: P_d versus ψ_T for the AMF-RAO with $N = 16$, $K = 32$, $P_{fa} = 10^{-4}$, and SNR= 19 dB.

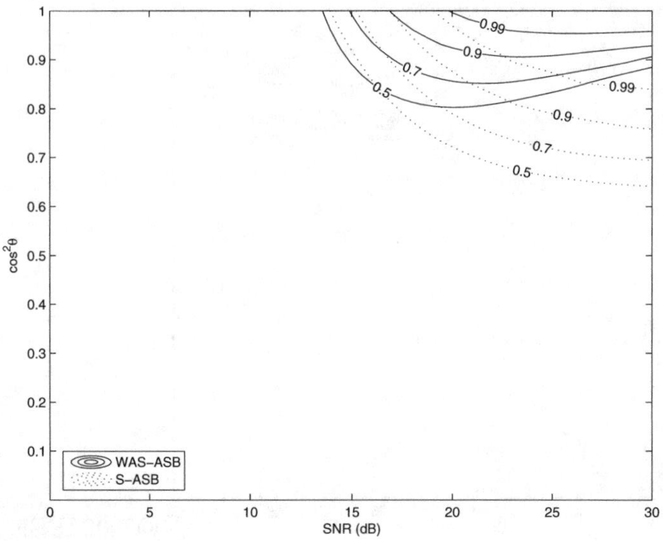

Figure 4.24: Contours of constant P_d for the WAS-ASB and the S-ASB with $N = 16$, $K = 32$, $r = 2$, and threshold pairs corresponding to the most selective cases of Figures 4.17 and 4.22.

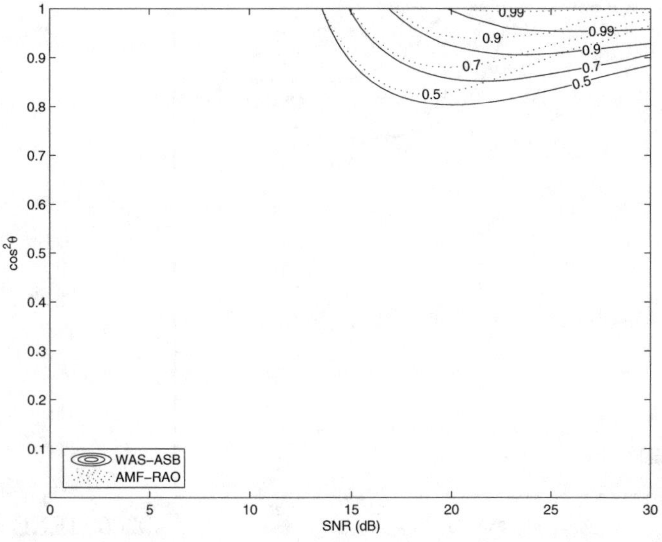

Figure 4.25: Contours of constant P_d for the WAS-ASB and the AMF-RAO with $N = 16$, $K = 32$, $r = 2$, and threshold pairs corresponding to the most selective cases of Figures 4.22 and 4.23.

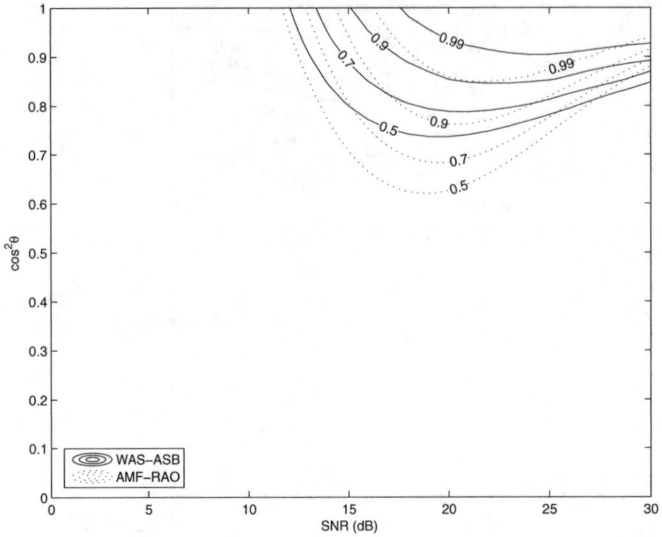

Figure 4.26: Contours of constant P_d for the WAS-ASB and the AMF-RAO with $N = 16$, $K = 48$, $r = 2$, and threshold pairs corresponding to the most selective cases for $P_{fa} = 10^{-4}$.

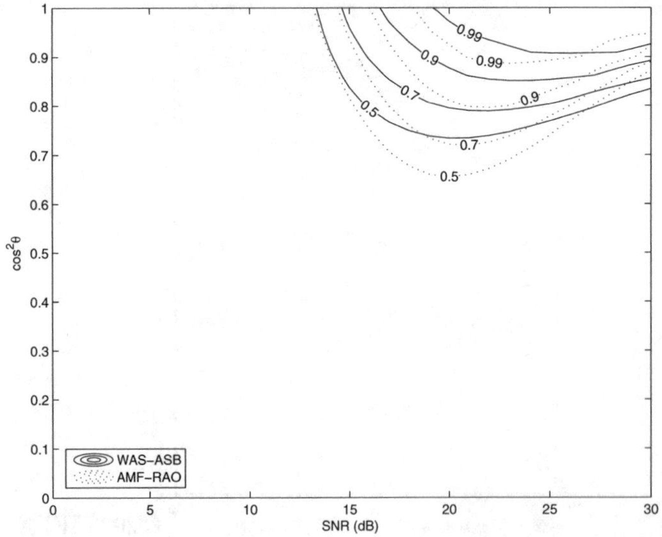

Figure 4.27: Contours of constant P_d for the WAS-ASB and the AMF-RAO with $N = 30$, $K = 60$, $r = 2$, and threshold pairs corresponding to the most selective cases for $P_{fa} = 10^{-4}$.

that the WAS-ASB is slightly superior to the AMF-RAO in terms of selectivity for the considered parameters values.

As a final analysis, we present some results relative to a case study which is quite typical in radar scenarios, i.e., the case where the disturbance entering the radar is composed by background (white) noise plus a certain number of noise like interferers with a very high power level (it is understood that this disturbance is present in all of the cells under consideration including the CUT). Specifically, we model the noise vectors according to the following covariance matrix [3]

$$\boldsymbol{R} = \sigma_n^2 \boldsymbol{I} + \frac{1}{N} \sum_{i=1}^{m} \sigma_i^2 \boldsymbol{a}(\psi_i) \boldsymbol{a}^{\dagger}(\psi_i), \tag{4.10}$$

where σ_n^2 is the power of the background noise, m is the number of noise like interferers, $\boldsymbol{a}(\psi)$ is given by equation (3.10), with σ_i^2 and ψ_i denoting the power and the angle of arrival of the i-th interferer, respectively. In particular, we assume $m = 2$, $\sigma_n^2 = 1$, interferers with the same power, interference-to-white-noise ratio (IWNR), namely

$$\text{IWNR} = \frac{\sigma_i^2}{\sigma_n^2},$$

equal to 40 dB,

$$[\psi_1 \ \psi_2] = [\pi/2 - 20\pi/360 \ \ \pi/2 - 40\pi/360].$$

As to $\boldsymbol{p}$, we assume the same model used throughout this book, namely $\boldsymbol{p} = \boldsymbol{a}(\pi/2 - \psi_T)$; this coherent signal has to be considered either as a useful target echo or as a sidelobe interferer according to the preassigned application requirements. In Figures 4.28-4.30 we plot P_d vs ψ_T (measured in degrees) for the ASB, the WAS-ASB, and the AMF-RAO, respectively, assuming $N = 16$, $K = 32$, $P_{fa} = 10^{-4}$, and SNR $= 19$ dB. Inspection of the figures indicates that this new scenario does not affect the relative ranking, in terms of tunability, of the three detectors (for a comparison see Figures 4.16, 4.22, and 4.23). Note though the null of the WAS-ASB at 10° corresponding to the direction of arrival of one of the two noise like interferers. Finally, in Figures 4.31 and 4.32, we present results of P_d vs ψ_T for the WAS-ASB and the AMF-RAO for $N = 30$, $K = 60$, SNR $= 20$ dB, and remaining parameters as in Figures 4.29 and 4.30. As already shown for mismatched signals in exponentially-correlated noise vectors, the WAS-ASB exhibits a wider tunability range than the AMF-RAO.

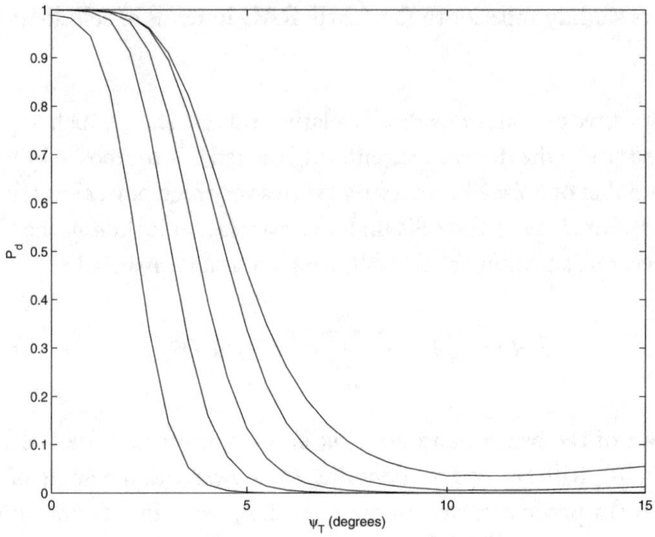

Figure 4.28: P_d versus ψ_T for the ASB with $N = 16$, $K = 32$, $P_{fa} = 10^{-4}$, SNR= 19 dB, and in presence of two noise like interferers (direction of arrival 10° and 20° off the nominal one), IWNR= 40 dB.

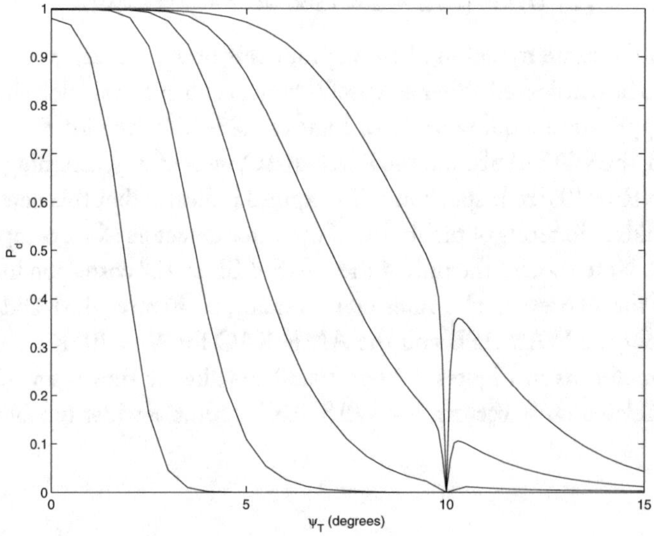

Figure 4.29: P_d versus ψ_T for the WAS-ASB with $N = 16$, $K = 32$, $r = 2$, $P_{fa} = 10^{-4}$, SNR= 19 dB, and in presence of two noise like interferers (direction of arrival 10° and 20° off the nominal one), IWNR= 40 dB.

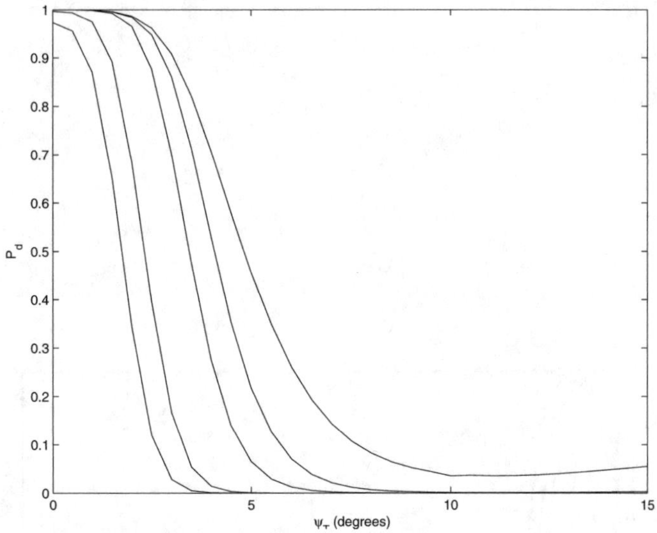

Figure 4.30: P_d versus ψ_T for the AMF-RAO with $N = 16$, $K = 32$, $P_{fa} = 10^{-4}$, SNR= 19 dB, and in presence of two noise like interferers (direction of arrival 10° and 20° off the nominal one), IWNR= 40 dB.

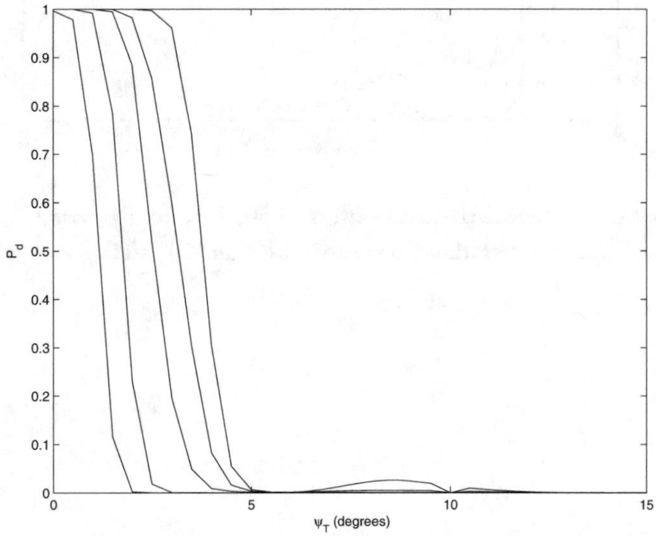

Figure 4.31: P_d versus ψ_T for the WAS-ASB with $N = 30$, $K = 60$, $r = 2$, $P_{fa} = 10^{-4}$, SNR= 20 dB, and in presence of two noise like interferers (direction of arrival 10° and 20° off the nominal one), IWNR= 40 dB.

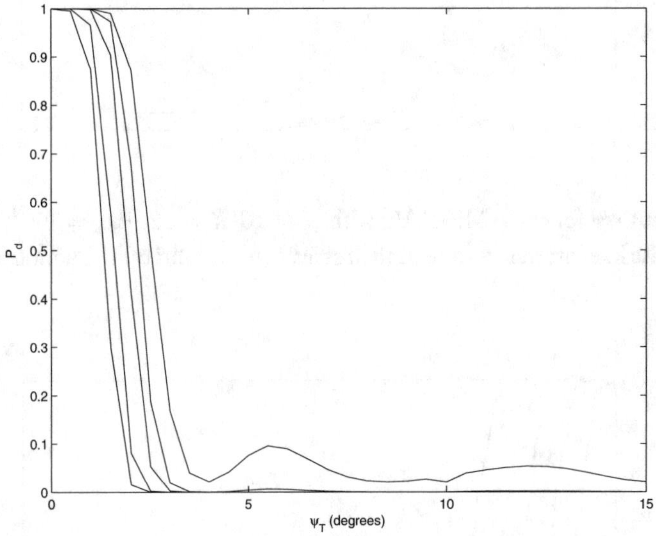

Figure 4.32: P_d versus ψ_T for the AMF-RAO with $N = 30$, $K = 60$, $P_{fa} = 10^{-4}$, SNR= 20 dB, and in presence of two noise like interferers (direction of arrival 10° and 20° off the nominal one), IWNR= 40 dB.

CHAPTER 5
Conclusions

This book has addressed the design and the analysis of adaptive decision schemes capable of trading detection performance of slightly mismatched signals for rejection capabilities of sidelobe interferers. The emphasis is on two-stage CFAR receivers: it has been shown that the WAS-ASB and the AMF-RAO can outperform the well-known ASB in terms of robustness and/or selectivity. In fact, the WAS-ASB is more robust than the ASB and the AMF-RAO; moreover, the WAS-ASB and the AMF-RAO are more selective than the ASB. The performance assessment has been conducted analytically for both matched and mismatched signals.

Several important issues would deserve further attention. First, observe that the WAS-ASB does not share with the ASB and the AMF-RAO the invariance to the group of transformations given in [47]; moreover, the WAS-ASB is slightly more time consuming than the ASB and the AMF-RAO. These differences suggest to compare the AMF-RAO and a two-stage detector obtained by cascading the W-ABORT to the AMF, but also to compare a subspace detector followed by the Rao test to the WAS-ASB (since in both cases the considered detectors would share the same invariance and complexity).

From a more practical point of view, it is of interest to consider the presence of dominant noise like interferers. For this case, detectors resorting to diagonal loading, although not strictly CFAR, may be preferred to GLRT-based and other ad hoc detectors [48, and references therein]. However, the behavior of diagonally loaded solutions has not been adequately investigated in terms of selectivity. Thus, another open issue is the design and the analysis of diagonally-loaded solutions retaining the selective behavior of WAS-ASB and AMF-RAO.

Another key point that has not been sufficiently investigated is the design and the analysis of optimized detection strategies for extended and/or distributed targets; a first attempt towards the design of tunable schemes to detect distributed targets has been considered in [49] resorting to the cone idea.

Since the Gaussian assumption is not always met, it would also be important to assess the performance of the considered detectors in non-Gaussian disturbance. A very preliminary study has been conducted for the S-ASB assuming a clutter-dominated environment modeled in terms of a compound-Gaussian process [13]. The analysis has shown that the CFAR property is not generally guaranteed but for the limiting case of one and the same texture component. Moreover, in the (limiting) case of cell-to-cell independent texture the performance assessment, conducted by Monte Carlo simulation, showed that the S-ASB, although less sensitive than the ASB to the "level of non-Gaussianity," may suffer an unacceptable inflation of false alarms. Finally, even a relatively moderate non-Gaussianity produces smoother transitions between "bandpass" and "stopband" regions, thus limiting the capacity of the S-ASB to discriminate between (what is classified as) a mainlobe target and a sidelobe interferer. Summarizing, a definite validation of the proposed strategies should take into account the possible non-Gaussian nature of the disturbance. Heterogeneity should also be considered, at least at the analysis stage, resorting to real data recordings. This analysis would probably indicate that such more adverse scenarios require different design solutions.

Another issue that has been considered, see for instance [50], but that could deserve further attention is that of classification of coherent signals as mainlobe or sidelobe signals. As a matter of fact, although proper selection of the thresholds of a two-stage receiver already allows to adjust directivity and, eventually, correctly classify mainlobe and sidelobe signals, the problem could be restated as a ternary hypothesis test [51] and solved as such.

APPENDIX A

Complex Normal Related Statistics

In this appendix we provide a short review of the multivariate complex normal distribution and of distributions related to it. Most of the results reported here are derived in [9, 20].

Consider a complex Gaussian vector, namely a random vector

$$z = x + jy \in \mathbb{C}^{N\times 1}, \quad x, y \in \mathbb{R}^{N\times 1},$$

where x and y are such that $w = [x^T \; y^T]^T$ is a normal vector with mean vector $\mu_w = [\mu_x^T \; \mu_y^T]^T$ and covariance matrix

$$\Sigma_{ww} = \begin{bmatrix} E[(x-\mu_x)(x-\mu_x)^T] & E[(x-\mu_x)(y-\mu_y)^T] \\ E[(y-\mu_y)(x-\mu_x)^T] & E[(y-\mu_y)(y-\mu_y)^T] \end{bmatrix} = \begin{bmatrix} \Sigma_{xx} & \Sigma_{xy} \\ \Sigma_{yx} & \Sigma_{yy} \end{bmatrix}.$$

As a preliminary step toward the definition of the complex normal distribution, observe the following facts.

Proposition A.1. *Assume that*

$$\Sigma_{xx} = \Sigma_{yy}, \tag{A.1}$$
$$\Sigma_{yx} = -\Sigma_{xy}. \tag{A.2}$$

It follows that the (symmetric and positive semidefinite) matrix

$$\Sigma_{ww} = \begin{bmatrix} \Sigma_{xx} & \Sigma_{xy} \\ -\Sigma_{xy} & \Sigma_{xx} \end{bmatrix}$$

is nonsingular if the (symmetric and positive semidefinite) matrix Σ_{xx} is nonsingular. Moreover, the determinant of the (Hermitian and positive semidefinite) matrix

$$\Sigma_{zz} = E[(z - E[z])(z - E[z])^\dagger] = 2(\Sigma_{xx} - j\Sigma_{xy}) \tag{A.3}$$

can be rewritten as

$$\det(\Sigma_{zz}) = 2^N \sqrt{\det(\Sigma_{ww})}, \tag{A.4}$$

where $\sqrt{(\cdot)}$ denotes the nonnegative square root of its argument. In particular, Σ_{zz} is nonsingular if and only if Σ_{ww} is nonsingular.

Proof. First observe that a nonsingular (and, hence, positive definite) $\boldsymbol{\Sigma}_{xx}$ implies (see formula (6) of paragraph 4.2.2 in [52])

$$\begin{aligned}\det(\boldsymbol{\Sigma}_{ww}) &= \det(\boldsymbol{\Sigma}_{xx})\det\left(\boldsymbol{\Sigma}_{xx}+\boldsymbol{\Sigma}_{xy}\boldsymbol{\Sigma}_{xx}^{-1}\boldsymbol{\Sigma}_{xy}\right)\\ &= [\det(\boldsymbol{\Sigma}_{xx})]^2\det\left(\boldsymbol{I}+\boldsymbol{\Sigma}_{xx}^{-1/2}\boldsymbol{\Sigma}_{xy}\boldsymbol{\Sigma}_{xx}^{-1}\boldsymbol{\Sigma}_{xy}\boldsymbol{\Sigma}_{xx}^{-1/2}\right),\end{aligned}$$

where $\boldsymbol{\Sigma}_{xx}^{-1/2}$ denotes the positive definite square root[1] of $\boldsymbol{\Sigma}_{xx}^{-1}$. Since the matrix

$$\boldsymbol{I}+\boldsymbol{\Sigma}_{xx}^{-1/2}\boldsymbol{\Sigma}_{xy}\boldsymbol{\Sigma}_{xx}^{-1}\boldsymbol{\Sigma}_{xy}\boldsymbol{\Sigma}_{xx}^{-1/2}$$

is positive definite, it follows that $\det(\boldsymbol{\Sigma}_{ww}) > 0$.

In order to prove the second part of the proposition, it is sufficient to show that

$$2^{2N}\det(\boldsymbol{\Sigma}_{ww}) = \left[\det(\boldsymbol{\Sigma}_{zz})\right]^2.$$

To this end, let $\lambda_1, \ldots, \lambda_r$ be the r distinct eigenvalues of $\boldsymbol{\Sigma}_{zz}$, with multiplicity $k_1, \ldots, k_r$, respectively, and

$$\boldsymbol{a}_{i,1}+j\boldsymbol{b}_{i,1},\ \ldots,\ \boldsymbol{a}_{i,k_i}+j\boldsymbol{b}_{i,k_i},\quad \boldsymbol{a}_{i,h},\boldsymbol{b}_{i,h}\in\mathbb{R},\quad h=1,\ldots,k_i,$$

k_i orthogonal eigenvectors of $\boldsymbol{\Sigma}_{zz}$ corresponding to λ_i. By exploiting the definition of eigenvalue we obtain the following conditions

$$\boldsymbol{\Sigma}_{zz}(\boldsymbol{a}_{i,h}+j\boldsymbol{b}_{i,h}) = \lambda_i(\boldsymbol{a}_{i,h}+j\boldsymbol{b}_{i,h}) \Rightarrow \begin{cases}\boldsymbol{\Sigma}_{xx}\boldsymbol{a}_{i,h}+\boldsymbol{\Sigma}_{xy}\boldsymbol{b}_{i,h} = \frac{\lambda_i}{2}\boldsymbol{a}_{i,h}\\ \boldsymbol{\Sigma}_{xx}\boldsymbol{b}_{i,h}-\boldsymbol{\Sigma}_{xy}\boldsymbol{a}_{i,h} = \frac{\lambda_i}{2}\boldsymbol{b}_{i,h}\end{cases},\qquad h=1,\ldots,k_i.$$

It follows that

$$\begin{bmatrix}\boldsymbol{a}_{i,h}\\ \boldsymbol{b}_{i,h}\end{bmatrix} \text{ and } \begin{bmatrix}-\boldsymbol{b}_{i,h}\\ \boldsymbol{a}_{i,h}\end{bmatrix},\qquad h=1,\ldots,k_i,$$

are $2k_i$ orthogonal[2] eigenvectors of $\boldsymbol{\Sigma}_{ww}$ corresponding to the eigenvalue $\lambda_i/2$. Finally, it is sufficient to observe that

$$\det(\boldsymbol{\Sigma}_{ww}) = \prod_{i=1}^{r}\left(\frac{\lambda_i}{2}\right)^{2k_i} = \left(\prod_{i=1}^{r}\lambda_i^{k_i}\right)^2\frac{1}{2^{2N}} = \left[\det(\boldsymbol{\Sigma}_{zz})\right]^2\frac{1}{2^{2N}},$$

where we have also used the identity $\sum_{i=1}^{r} k_i = N$. □

[1]The square root of a positive definite matrix $\boldsymbol{A}\in\mathbb{C}^{N\times N}$ is given by [53]

$$\boldsymbol{A}^{1/2} = \boldsymbol{U}\boldsymbol{\Lambda}^{1/2}\boldsymbol{U}^\dagger,$$

where $\boldsymbol{U}$ is a unitary matrix whose columns are the orthonormal eigenvectors of $\boldsymbol{A}$ and $\boldsymbol{\Lambda}$ is a diagonal matrix whose diagonal entries are the corresponding eigenvalues. Moreover, $\boldsymbol{\Lambda}^{1/2}$ has to be intended as a diagonal matrix whose entries are the positive square roots of the entries of $\boldsymbol{\Lambda}$.

[2]Recall that $(\boldsymbol{a}_{i,h}+j\boldsymbol{b}_{i,h})^\dagger(\boldsymbol{a}_{i,l}+j\boldsymbol{b}_{i,l}) = 0, \forall h \neq l$.

Observe that if $\boldsymbol{\Sigma}_{xx}$ is nonsingular and conditions (A.1)-(A.2) are verified, $\boldsymbol{\Sigma}_{zz}$ and $\boldsymbol{\Sigma}_{ww}$ are non singular matrices too; hereafter we assume that $\boldsymbol{\Sigma}_{xx}$ is nonsingular.

Proposition A.2. *Assume that conditions* (A.1) *and* (A.2) *hold true. It follows that*

$$\boldsymbol{\Sigma}_{ww}^{-1} = \begin{bmatrix} \boldsymbol{\Sigma}_{xx} & \boldsymbol{\Sigma}_{xy} \\ -\boldsymbol{\Sigma}_{xy} & \boldsymbol{\Sigma}_{xx} \end{bmatrix}^{-1} = \begin{bmatrix} \boldsymbol{C}_{xx} & \boldsymbol{C}_{xy} \\ -\boldsymbol{C}_{xy} & \boldsymbol{C}_{xx} \end{bmatrix},$$

with

$$\begin{aligned} \boldsymbol{C}_{xx} &= (\boldsymbol{\Sigma}_{xx} + \boldsymbol{\Sigma}_{xy}\boldsymbol{\Sigma}_{xx}^{-1}\boldsymbol{\Sigma}_{xy})^{-1} \\ &= \boldsymbol{\Sigma}_{xx}^{-1} - \boldsymbol{\Sigma}_{xx}^{-1}\boldsymbol{\Sigma}_{xy}(\boldsymbol{\Sigma}_{xx} + \boldsymbol{\Sigma}_{xy}\boldsymbol{\Sigma}_{xx}^{-1}\boldsymbol{\Sigma}_{xy})^{-1}\boldsymbol{\Sigma}_{xy}\boldsymbol{\Sigma}_{xx}^{-1} \end{aligned}$$

and

$$\begin{aligned} \boldsymbol{C}_{xy} &= -\boldsymbol{\Sigma}_{xx}^{-1}\boldsymbol{\Sigma}_{xy}(\boldsymbol{\Sigma}_{xx} + \boldsymbol{\Sigma}_{xy}\boldsymbol{\Sigma}_{xx}^{-1}\boldsymbol{\Sigma}_{xy})^{-1} \\ &= -(\boldsymbol{\Sigma}_{xx} + \boldsymbol{\Sigma}_{xy}\boldsymbol{\Sigma}_{xx}^{-1}\boldsymbol{\Sigma}_{xy})^{-1}\boldsymbol{\Sigma}_{xy}\boldsymbol{\Sigma}_{xx}^{-1}. \end{aligned}$$

In particular, $\boldsymbol{C}_{xy}$ *is skew-symmetric and, hence,*

$$\boldsymbol{u}^T\boldsymbol{C}_{xy}\boldsymbol{u} = 0, \quad \forall \boldsymbol{u} \in \mathbb{R}^{N\times 1}.$$

Proof. The proof is a straightforward consequence of formulas (1) and (2) of paragraph 3.5.3 and formula (2) of paragraph 3.5.2 in [52]. □

Definition A.3. The complex Gaussian vector $z = \boldsymbol{x} + j\boldsymbol{y} \in \mathbb{C}^{N\times 1}$ is said to have a complex normal distribution with mean vector

$$\boldsymbol{\mu}_z = \boldsymbol{\mu}_x + j\boldsymbol{\mu}_y$$

and covariance matrix $\boldsymbol{\Sigma}_{zz}$, given by equation (A.3), if conditions (A.1) and (A.2) hold true. For such a vector we write $z \sim \mathcal{CN}_N(\boldsymbol{\mu}_z, \boldsymbol{\Sigma}_{zz})$.

It is possible to show that the pdf of z (equivalently of $\boldsymbol{w}$) admits the following representation [20].

Theorem A.4. *Let* $z \sim \mathcal{CN}_N(\boldsymbol{\mu}_z, \boldsymbol{\Sigma}_{zz})$*. The pdf of* z *can be recast as*

$$f_z(\boldsymbol{w}) = \frac{1}{\pi^N \det(\boldsymbol{\Sigma}_{zz})} e^{-(z-\boldsymbol{\mu}_z)^\dagger \boldsymbol{\Sigma}_{zz}^{-1}(z-\boldsymbol{\mu}_z)},$$

with $\boldsymbol{w} = [\boldsymbol{x}^T\ \boldsymbol{y}^T]^T$ *and* $z = \boldsymbol{x} + j\boldsymbol{y}$.

Proof. Observe that the pdf of z has the following expression

$$f_z(\boldsymbol{w}) = \frac{1}{(2\pi)^N \det^{1/2}(\boldsymbol{\Sigma}_{ww})} e^{-\frac{1}{2}(\boldsymbol{w}-\boldsymbol{\mu}_w)^T \boldsymbol{\Sigma}_{ww}^{-1}(\boldsymbol{w}-\boldsymbol{\mu}_w)}, \tag{A.5}$$

where $\boldsymbol{\mu}_w = [\boldsymbol{\mu}_x^T \ \boldsymbol{\mu}_y^T]^T$ and

$$\boldsymbol{\Sigma}_{ww} = \begin{bmatrix} \boldsymbol{\Sigma}_{xx} & \boldsymbol{\Sigma}_{xy} \\ -\boldsymbol{\Sigma}_{xy} & \boldsymbol{\Sigma}_{xx} \end{bmatrix}.$$

Resorting to Proposition A.2 the exponent of (A.5) can be rewritten as

$$\begin{aligned} &\frac{1}{2}(\boldsymbol{w} - \boldsymbol{\mu}_w)^T \boldsymbol{\Sigma}_{ww}^{-1}(\boldsymbol{w} - \boldsymbol{\mu}_w) \\ &= \frac{1}{2}[(\boldsymbol{x} - \boldsymbol{\mu}_x)^T \ (\boldsymbol{y} - \boldsymbol{\mu}_y)^T] \begin{bmatrix} \boldsymbol{C}_{xx} & \boldsymbol{C}_{xy} \\ -\boldsymbol{C}_{xy} & \boldsymbol{C}_{xx} \end{bmatrix} \begin{bmatrix} \boldsymbol{x} - \boldsymbol{\mu}_x \\ \boldsymbol{y} - \boldsymbol{\mu}_y \end{bmatrix} \\ &= (\boldsymbol{z} - \boldsymbol{\mu}_z)^\dagger \boldsymbol{A}(\boldsymbol{z} - \boldsymbol{\mu}_z), \end{aligned}$$

where $\boldsymbol{A} = \frac{1}{2}(\boldsymbol{C}_{xx} - j\boldsymbol{C}_{xy})$. Moreover, the same proposition allows to prove that

$$\boldsymbol{\Sigma}_{zz}\boldsymbol{A} = \boldsymbol{A}\boldsymbol{\Sigma}_{zz} = \boldsymbol{I},$$

namely that $\boldsymbol{A} = \boldsymbol{\Sigma}_{zz}^{-1}$. Use of identity (A.4) concludes the proof. □

Complex normal distributions have a central role in this work. Note also that, if $n(t)$ is a zero-mean complex Gaussian random process, which possesses the circular property, i.e., $n(t)$ and $n(t)e^{j\vartheta}$ are identically distributed for all $\vartheta \in [0,\ 2\pi)$, the vector

$$\boldsymbol{n} = \boldsymbol{n}_x + j\boldsymbol{n}_y, \quad \boldsymbol{n}_x, \boldsymbol{n}_y \in \mathbb{R}^{N\times 1},$$

obtained by collecting N samples of $n(t)$, is complex normal with zero mean. Indeed, since $n(t)$ and $jn(t)$ are identically distributed, the vectors

$$\begin{bmatrix} \boldsymbol{n}_x \\ \boldsymbol{n}_y \end{bmatrix} \text{ and } \begin{bmatrix} -\boldsymbol{n}_y \\ \boldsymbol{n}_x \end{bmatrix}$$

have the same covariance matrix, i.e.,

$$\begin{aligned} &E\left\{ \begin{bmatrix} \boldsymbol{n}_x \\ \boldsymbol{n}_y \end{bmatrix} [\boldsymbol{n}_x^T \ \boldsymbol{n}_y^T] \right\} = \begin{bmatrix} E\{\boldsymbol{n}_x\boldsymbol{n}_x^T\} & E\{\boldsymbol{n}_x\boldsymbol{n}_y^T\} \\ E\{\boldsymbol{n}_y\boldsymbol{n}_x^T\} & E\{\boldsymbol{n}_y\boldsymbol{n}_y^T\} \end{bmatrix} \\ &= E\left\{ \begin{bmatrix} -\boldsymbol{n}_y \\ \boldsymbol{n}_x \end{bmatrix} [-\boldsymbol{n}_y^T \ \boldsymbol{n}_x^T] \right\} = \begin{bmatrix} E\{\boldsymbol{n}_y\boldsymbol{n}_y^T\} & -E\{\boldsymbol{n}_y\boldsymbol{n}_x^T\} \\ -E\{\boldsymbol{n}_x\boldsymbol{n}_y^T\} & E\{\boldsymbol{n}_x\boldsymbol{n}_x^T\} \end{bmatrix} \end{aligned}$$

which, in turn, implies conditions (A.1) and (A.2) and, hence, that $\boldsymbol{n}$ obeys the complex normal distribution.

In the following we give definitions and results concerning statistics related to the complex normal distribution that are useful in the body of the book.

Proposition A.5. ***Complex Chi-Square Distribution*** *The random variable* $a = \|\boldsymbol{r}\|^2$, *with* $\boldsymbol{r} \sim \mathcal{CN}_N(\boldsymbol{0}, \boldsymbol{I})$, *is said*[3] *complex central chi-square with* N *complex degrees of freedom and its pdf is given by*

$$f_a(x) = \frac{x^{N-1}}{(N-1)!} e^{-x}, \quad x \geq 0.$$

We denote this as $a \sim \mathcal{C}\chi^2_N$.

More generally, the random variable $a = \|\boldsymbol{r}\|^2$, *with* $\boldsymbol{r} \sim \mathcal{CN}_N(\boldsymbol{s}, \boldsymbol{I})$, *is said complex noncentral chi-square with* N *complex degrees of freedom and noncentrality parameter* δ, *with* $\delta^2 = \|\boldsymbol{s}\|^2$. *The pdf of* a *is*

$$f_a(x) = \frac{x^{N-1}}{(N-1)!} e^{-x-\delta^2} {}_0F_1(N; \delta^2 x), \quad x \geq 0,$$

where ${}_0F_1$ *is a hypergeometric function* [54]. *We write* $a \sim \mathcal{C}\chi^2_N(\delta)$.

Proposition A.6. ***Complex F-distribution*** *The ratio* $c = a/b$, *where* $a \sim \mathcal{C}\chi^2_N$ *is independent of* $b \sim \mathcal{C}\chi^2_M$, *obeys the complex central F-distribution with* N, M *complex degrees of freedom. The pdf of* c *is given by*

$$f_c(x) = \frac{(N+M-1)!}{(N-1)!(M-1)!} \frac{x^{N-1}}{(1+x)^{N+M}}, \quad x \geq 0,$$

while the CDF is

$$F_c(x) = \frac{x^N}{(1+x)^{N+M-1}} \sum_{k=0}^{M-1} \binom{N+M-1}{N+k} x^k, \quad x \geq 0.$$

We denote this as $c \sim \mathcal{CF}_{N,M}$.

More generally, if $a \sim \mathcal{C}\chi^2_N(\delta)$ *and* $b \sim \mathcal{C}\chi^2_M$ *independent of* a, *then* $c = a/b$ *is ruled by the complex noncentral F-distribution with* N, M *complex degrees of freedom and noncentrality parameter* δ. *The resulting pdf and CDF are given by* [9]

$$f_c(x) = \frac{(N+M-1)!}{(N-1)!(M-1)!} \frac{x^{N-1}}{(1+x)^{N+M}} e^{-\delta^2/(1+x)} \times \sum_{k=0}^{M} \binom{M}{k} \frac{(N-1)!}{(N+k-1)!} \left(\frac{\delta^2 x}{1+x}\right)^k, \quad x \geq 0,$$

[3]Remember that $\boldsymbol{0}$ denotes a matrix with all zero entries, $\boldsymbol{I}$ an identity matrix, and $\|\cdot\|$ is the Euclidean norm of a vector.

and

$$F_c(x) = \frac{x^N}{(1+x)^{N+M-1}} \sum_{k=0}^{M-1} \binom{N+M-1}{N+k} x^k e^{-\delta^2/(1+x)} \sum_{i=0}^{k} \left(\frac{\delta^2}{1+x}\right)^i \frac{1}{i!}, \quad x \geq 0,$$

respectively. We write $c \sim \mathcal{CF}_{N,M}(\delta)$.

Remark. Let $c \sim \mathcal{CF}_{N,M}$, it follows that

$$\mathrm{P}[c > x] = 1 - \mathrm{P}[c \leq x] = 1 - \frac{x^N}{(1+x)^{N+M-1}} \sum_{k=0}^{M-1} \binom{N+M-1}{N+k} x^k, \quad x \geq 0. \tag{A.6}$$

On the other hand, substituting $N+k$ with l in (A.6) yields

$$\begin{aligned}
\mathrm{P}[c > x] &= 1 - \frac{1}{(1+x)^{N+M-1}} \sum_{l=N}^{N+M-1} \binom{N+M-1}{l} x^l \\
&= 1 - \frac{1}{(1+x)^{N+M-1}} \left\{ \sum_{l=N}^{N+M-1} \binom{N+M-1}{l} x^l + \sum_{l=0}^{N-1} \binom{N+M-1}{l} x^l \right. \\
&\quad \left. - \sum_{l=0}^{N-1} \binom{N+M-1}{l} x^l \right\} \\
&= \frac{1}{(1+x)^{N+M-1}} \sum_{l=0}^{N-1} \binom{N+M-1}{l} x^l, \quad x \geq 0.
\end{aligned} \tag{A.7}$$

The above equation is useful to derive closed-form expressions for the P_{fa} of the receivers analyzed in this book.

Proposition A.7. ***Complex Beta Distribution*** *Let* $a \sim \mathcal{CF}_{N,M}$, *then*

$$b = \frac{1}{1+a}$$

obeys the complex central beta distribution with M, N *complex degrees of freedom. We denote this as* $b \sim \mathcal{C}\beta_{M,N}$. *The pdf and CDF of b are*

$$f_b(x) = \frac{(N+M-1)!}{(N-1)!(M-1)!} x^{M-1} (1-x)^{N-1}, \quad 0 \leq x \leq 1,$$

and

$$F_b(x) = x^{N+M-1} \sum_{k=0}^{N-1} \binom{N+M-1}{k} \left(\frac{1-x}{x}\right)^k, \quad 0 \leq x \leq 1,$$

respectively.

More generally, if $a \sim \mathcal{CF}_{N,M}(\delta)$, *then* $b = 1/(1+a)$ *is ruled by the complex noncentral beta distribution with* M, N *complex degrees of freedom and noncentrality parameter* δ. *The pdf of such a random variable is given by*

$$f_b(x) = e^{-\delta^2 x} \sum_{k=0}^{M} \binom{M}{k} \frac{(N+M-1)!}{(N+k-1)!(M-1)!} \delta^{2k} x^{M-1} (1-x)^{N+k-1}, \quad 0 \le x \le 1,$$

while the CDF is

$$F_b(x) = 1 - x^{M-1}(1-x)^N \sum_{k=0}^{M-1} \binom{N+M-1}{k+N} \left(\frac{1-x}{x}\right)^k \times e^{-\delta^2 x} \sum_{i=0}^{k} \frac{(\delta^2 x)^i}{i!}, \quad 0 \le x \le 1.$$

We write $b \sim \mathcal{C}\beta_{M,N}(\delta)$.

Definition A.8. Complex Wishart Distribution. Let

$$\boldsymbol{Z} = [z_1 \cdots z_K] \in \mathbb{C}^{N \times K}, \text{ with } z_i \sim \mathcal{CN}_N(\boldsymbol{0}, \boldsymbol{M}), \ i = 1, \dots, K,$$

independent (and identically distributed) random vectors. Then, the random matrix

$$\boldsymbol{S} = \boldsymbol{Z}\boldsymbol{Z}^\dagger = \sum_{i=1}^{K} z_i z_i^\dagger \in \mathbb{C}^{N \times N}$$

is said to have the complex central Wishart distribution with parameters K and $\boldsymbol{M}$ and we write $\boldsymbol{S} \sim \mathcal{CW}_N(K, \boldsymbol{M})$ where the subscript on $\mathcal{CW}$ denotes the size of the matrix $\boldsymbol{S}$.

Two important properties of the Wishart distribution are given below without proof [33].

Proposition A.9. ***Property 1 of the Wishart Distribution*** *Let* $\boldsymbol{A} \in \mathbb{C}^{N \times P}$ *a matrix of rank* P *and* $\boldsymbol{S} \sim \mathcal{CW}_N(K, \boldsymbol{M})$, *with* $\boldsymbol{M}$ *a positive definite Hermitian matrix, then*

$$\boldsymbol{A}^\dagger \boldsymbol{S} \boldsymbol{A} \sim \mathcal{CW}_P(K, \boldsymbol{A}^\dagger \boldsymbol{M} \boldsymbol{A}).$$

It follows that $\boldsymbol{S} \sim \mathcal{CW}_N(K, \boldsymbol{M})$ implies

$$\boldsymbol{M}^{-1/2} \boldsymbol{S} \boldsymbol{M}^{-1/2} \sim \mathcal{CW}_N(K, \boldsymbol{I}),$$

where $\boldsymbol{M}^{-1/2}$ is the (positive definite) square root of $\boldsymbol{M}^{-1}$.

Proposition A.10. ***Property 2 of the Wishart Distribution*** *Let* $\boldsymbol{S} \sim \mathcal{CW}_N(K, \boldsymbol{M})$, *with* $\boldsymbol{M}$ *a positive definite Hermitian matrix, then the random matrix* $\boldsymbol{S}$ *is positive definite with probability one if and only if* $K \geq N$.

We conclude this appendix with two theorems. The first one is an extension of Theorem 3.2.10 in [33] to the complex case, also reported in [9, pp. 152-155], while the second one is a straightforward generalization of Theorem 3.2.12 in [33], by means of the properties of the complex Wishart distribution given in [55, 56].

Theorem A.11. Partitioned Wishart Matrices. *Let* $\boldsymbol{S} \sim \mathcal{CW}_m(n, \boldsymbol{M})$, *with* $\boldsymbol{M}$ *a positive definite Hermitian matrix, be partitioned as follows*

$$\boldsymbol{S} = \begin{bmatrix} \underset{r,r}{\boldsymbol{S}_{11}} & \underset{r,m-r}{\boldsymbol{S}_{12}} \\ \underset{m-r,r}{\boldsymbol{S}_{21}} & \underset{m-r,m-r}{\boldsymbol{S}_{22}} \end{bmatrix},$$

with

$$\boldsymbol{M} = \begin{bmatrix} \underset{r,r}{\boldsymbol{M}_{11}} & \underset{r,m-r}{\boldsymbol{M}_{12}} \\ \underset{m-r,r}{\boldsymbol{M}_{21}} & \underset{m-r,m-r}{\boldsymbol{M}_{22}} \end{bmatrix}$$

and $n > m - r$. *Then,* $\boldsymbol{S}_{22}$ *is* $\mathcal{CW}_{m-r}(n, \boldsymbol{M}_{22})$ *and it is nonsingular with probability one. Now define*

$$\begin{aligned} \boldsymbol{A} &= \boldsymbol{S}_{11} - \boldsymbol{S}_{12}\boldsymbol{S}_{22}^{-1}\boldsymbol{S}_{21}, \\ \boldsymbol{B} &= \boldsymbol{M}_{11} - \boldsymbol{M}_{12}\boldsymbol{M}_{22}^{-1}\boldsymbol{M}_{21}. \end{aligned}$$

Then, $\boldsymbol{A}$ *is* $\mathcal{CW}_r(n - m + r, \boldsymbol{B})$ *and it is independent of the set* $\{\boldsymbol{S}_{12}, \boldsymbol{S}_{22}\}$.

Lemma A.12. *Let* $\boldsymbol{A} \in \mathbb{C}^{N \times P}$ *be a matrix of rank* P *and* $\boldsymbol{S} \sim \mathcal{CW}_N(K, \boldsymbol{M})$, *with* $\boldsymbol{M}$ *a positive definite Hermitian matrix and* $K \geq N$, *then*

$$(\boldsymbol{A}^\dagger \boldsymbol{S}^{-1} \boldsymbol{A})^{-1} \sim \mathcal{CW}_P(K - N + P, (\boldsymbol{A}^\dagger \boldsymbol{M}^{-1} \boldsymbol{A})^{-1}).$$

Theorem A.13. *If* $\boldsymbol{A} \sim \mathcal{CW}_m(n, \boldsymbol{\Sigma})$, $n \geq m$, *with* $\boldsymbol{\Sigma}$ *a positive definite Hermitian matrix, and* $\boldsymbol{y} \in \mathbb{C}^{m \times 1}$ *any random vector distributed independently of* $\boldsymbol{A}$ *with* $P(\boldsymbol{y} = \boldsymbol{0}) = 0$, *then*

$$\frac{\boldsymbol{y}^\dagger \boldsymbol{\Sigma}^{-1} \boldsymbol{y}}{\boldsymbol{y}^\dagger \boldsymbol{A}^{-1} \boldsymbol{y}}$$

is a complex central chi-square random variable, with $n - m + 1$ *degrees of freedom, independent of* $\boldsymbol{y}$.

Proof. By Lemma A.12 the random variable $(\boldsymbol{y}^\dagger \boldsymbol{A}^{-1}\boldsymbol{y})^{-1}$, given $\boldsymbol{y}$, is ruled by $\mathcal{CW}_1(n-m+1, (\boldsymbol{y}^\dagger \boldsymbol{\Sigma}^{-1}\boldsymbol{y})^{-1})$. By Proposition A.9 it follows that, given $\boldsymbol{y}$,

$$\frac{\boldsymbol{y}^\dagger \boldsymbol{\Sigma}^{-1}\boldsymbol{y}}{\boldsymbol{y}^\dagger \boldsymbol{A}^{-1}\boldsymbol{y}} \sim \mathcal{C}\chi^2_{n-m+1}.$$

Since the above distribution does not depend on $\boldsymbol{y}$, it is also the unconditional one. □

APPENDIX B

Statistical Characterization of Kelly's detector, AMF, ACE, and W-ABORT

First of all, rewrite Kelly's detector in the equivalent form

$$\tilde{t}_{\mathrm{K}} = \frac{t_{\mathrm{K}}}{1 - t_{\mathrm{K}}} = \frac{\boldsymbol{r}^\dagger \boldsymbol{S}^{-1} \boldsymbol{v} (\boldsymbol{v}^\dagger \boldsymbol{S}^{-1} \boldsymbol{v})^{-1} \boldsymbol{v}^\dagger \boldsymbol{S}^{-1} \boldsymbol{r}}{\left(1 + \boldsymbol{r}^\dagger \boldsymbol{S}^{-1} \boldsymbol{r} - \dfrac{|\boldsymbol{r}^\dagger \boldsymbol{S}^{-1} \boldsymbol{v}|^2}{\boldsymbol{v}^\dagger \boldsymbol{S}^{-1} \boldsymbol{v}}\right)} \tag{B.1}$$

and define β as follows

$$\beta = \frac{1}{1 + \boldsymbol{r}^\dagger \boldsymbol{S}^{-1} \boldsymbol{r} - \dfrac{|\boldsymbol{r}^\dagger \boldsymbol{S}^{-1} \boldsymbol{v}|^2}{\boldsymbol{v}^\dagger \boldsymbol{S}^{-1} \boldsymbol{v}}}.$$

Under the noise-only hypothesis, i.e.,

$$\boldsymbol{r} = \boldsymbol{n}, \quad \boldsymbol{n} \sim \mathcal{CN}_N(\boldsymbol{0}, \boldsymbol{M}),$$

the random variable $\tilde{t}_{\mathrm{K}}$ is distributed according to a complex central F-distribution with $1, K - N + 1$ degrees of freedom and it is independent of β, which, in turn, obeys a complex central beta distribution with $K - N + 2, N - 1$ degrees of freedom [2, 9]. By resorting to equation (A.7) and observing that the distribution of $\tilde{t}_{\mathrm{K}}$ does not depend on $\boldsymbol{M}$, the P_{fa} of Kelly's detector can be easily evaluated as follows

$$P_{fa}(\eta) = \mathrm{P}[t_{\mathrm{K}} > \eta; H_0] = \mathrm{P}[\tilde{t}_{\mathrm{K}} > \tilde{\eta}; H_0] = \frac{1}{(1 + \tilde{\eta})^{K-N+1}},$$

where $\tilde{\eta} = \eta/(1 - \eta)$.

On the other hand, under the H_1 hypothesis and assuming that the actual steering vector $\boldsymbol{p}$ may be different from the nominal one $\boldsymbol{v}$, i.e.,

$$\boldsymbol{r} = \alpha \boldsymbol{p} + \boldsymbol{n}, \quad \boldsymbol{n} \sim \mathcal{CN}_N(\boldsymbol{0}, \boldsymbol{M}),$$

$\tilde{t}_{\mathrm{K}}$, given β, is ruled by a complex noncentral F-distribution with $1, K - N + 1$ degrees of freedom and noncentrality parameter δ, with [37, 38]

$$\delta^2 = \mathrm{SNR}\, \beta \, \cos^2 \theta,$$

where

$$\mathrm{SNR} = |\alpha|^2 \, \boldsymbol{p}^\dagger \boldsymbol{M}^{-1} \boldsymbol{p} \tag{B.2}$$

and

$$\cos^2\theta = \frac{|\boldsymbol{p}^\dagger \boldsymbol{M}^{-1} \boldsymbol{v}|^2}{(\boldsymbol{v}^\dagger \boldsymbol{M}^{-1} \boldsymbol{v})(\boldsymbol{p}^\dagger \boldsymbol{M}^{-1} \boldsymbol{p})}.$$

In addition, the random variable β obeys the complex noncentral beta distribution with $K - N + 2$, $N - 1$ degrees of freedom and noncentrality parameter δ_β, with

$$\delta_\beta^2 = \mathrm{SNR}\ \sin^2\theta.$$

In the special case of perfect match between $\boldsymbol{p}$ and $\boldsymbol{v}$, we obtain

$$\begin{aligned} \delta^2 &= \mathrm{SNR}\,\beta, \\ \delta_\beta^2 &= 0, \end{aligned}$$

namely $\beta \sim \mathcal{C}\beta_{K-N+2,N-1}$ and $\tilde{t}_{\mathrm{K}} \sim \mathcal{CF}_{1,K-N+1}(\delta)$, given β. Note that since the random variable β belongs to $[0, 1]$, it tends to lower the SNR, i.e.,

$$\mathrm{SNR} \geq \mathrm{SNR}\beta, \quad \forall\, \beta \in [0, 1];$$

for this reason it is referred to as *loss factor*.

In the sequel, we express the statistics of AMF, ACE, and W-ABORT as functions of $\tilde{t}_{\mathrm{K}}$ and β and compute[1] the corresponding P_{fa} resorting to equation (A.7).

First, observe that an immediate consequence of equation (B.1) is the following

$$\tilde{t}_{\mathrm{K}} = t_{\mathrm{AMF}}\beta \Rightarrow t_{\mathrm{AMF}} = \frac{\tilde{t}_{\mathrm{K}}}{\beta}.$$

The P_{fa} of the AMF can be written as

$$P_{fa}(\eta) = \mathrm{P}[t_{\mathrm{AMF}} > \eta;\, H_0] = \frac{K!}{(K-N+1)!(N-2)!} \int_0^1 \frac{x^{K-N+1}(1-x)^{N-2}}{(1+\eta x)^{K-N+1}} dx\,.$$

As to the ACE, observe that

$$\begin{aligned} \tilde{t}_{\mathrm{ACE}} &= \frac{t_{\mathrm{ACE}}}{1 - t_{\mathrm{ACE}}} = \frac{|\boldsymbol{r}^\dagger \boldsymbol{S}^{-1} \boldsymbol{v}|^2}{(\boldsymbol{v}^\dagger \boldsymbol{S}^{-1} \boldsymbol{v})(\boldsymbol{r}^\dagger \boldsymbol{S}^{-1} \boldsymbol{r}) - |\boldsymbol{r}^\dagger \boldsymbol{S}^{-1} \boldsymbol{v}|^2} = \frac{t_{\mathrm{AMF}}}{\beta^{-1} - 1} \\ &= \frac{\tilde{t}_{\mathrm{K}}}{(1-\beta)}. \end{aligned}$$

[1]Remember that the distributions of $\tilde{t}_{\mathrm{K}}$ and β under H_0 do not depend on $\boldsymbol{M}$, hence the computation of the P_{fa}'s does not require evaluating suprema.

Thus, the P_{fa} of the ACE is given by

$$P_{fa}(\eta) = \mathrm{P}[t_{\text{ACE}} > \eta;\, H_0]$$
$$= \mathrm{P}[\tilde{t}_{\text{ACE}} > \tilde{\eta};\, H_0] = \frac{K!}{(K-N+1)!(N-2)!}\int_0^1 \frac{x^{K-N+1}(1-x)^{N-2}}{[1+\tilde{\eta}(1-x)]^{K-N+1}}dx,$$

where $\tilde{\eta} = \eta/(1-\eta)$.

As to the W-ABORT, we can rewrite equation (3.18) as

$$t_{\text{WA}} = \frac{1 + \boldsymbol{r}^{\dagger}\boldsymbol{S}^{-1}\boldsymbol{r} - t_{\text{AMF}} + t_{\text{AMF}}}{\beta^{-2}} = \frac{\beta^{-1} + \tilde{t}_{\text{K}}/\beta}{\beta^{-2}} = \beta(1+\tilde{t}_{\text{K}}).$$

Again, using equation (A.7) together with some algebra, it is possible to show that the P_{fa} of the W-ABORT, i.e., $P_{fa}(\eta) = \mathrm{P}[t_{\text{WA}} > \eta;\, H_0]$, has the following expressions

- $\eta \geq 1$

$$P_{fa}(\eta) = \frac{1}{\eta^{K-N+1}} \frac{K!}{(K-N+1)!\prod_{i=1}^{N-1}[2(K-N+1)+i]};$$

- $0 \leq \eta < 1$

$$P_{fa}(\eta) = \frac{K!}{(K-N+1)!(N-2)!}\sum_{l=0}^{N-2}\binom{N-2}{l}\frac{(-1)^{N-2-l}\eta^{K-l}}{2K-N+1-l} + 1 - A(\eta),$$

where

$$A(\eta) = \eta^K \sum_{l=0}^{N-2}\binom{K}{l}\left(\frac{1-\eta}{\eta}\right)^l.$$

Further details on the analysis of the W-ABORT can be found in [57]. The statistical characterization of the above detectors is summarized in Table B.1.

For the sake of completeness, we also consider the GLRT for known $\boldsymbol{M}$ which is given by

$$t_{\text{M}} = \frac{|\boldsymbol{r}^{\dagger}\boldsymbol{M}^{-1}\boldsymbol{v}|^2}{\boldsymbol{v}^{\dagger}\boldsymbol{M}^{-1}\boldsymbol{v}} \underset{H_0}{\overset{H_1}{\gtrless}} \eta.$$

It is easy to verify that under the H_0 hypothesis t_{M} obeys the complex central chi-square distribution with 1 degree of freedom. Consequently, the P_{fa} is given by

$$P_{fa}(\eta) = \mathrm{P}[t_{\text{M}} > \eta;\, H_0] = e^{-\eta}.$$

On the other hand, when the H_1 hypothesis is in force, t_{M} is ruled by the complex noncentral chi-square distribution with 1 degree of freedom and noncentrality parameter δ, with $\delta^2 = \text{SNR}$.

Table B.1: Distributions of Adaptive Detectors.

Decision Statistic	Statistical Characterization
Kelly's detector $\left(\tilde{t}_{\mathrm{K}} = \dfrac{t_{\mathrm{K}}}{1 - t_{\mathrm{K}}}\right)$	$H_0 : \tilde{t}_{\mathrm{K}} \sim \mathcal{CF}_{1,K-N+1}$ independent of β, $\beta \sim \mathcal{C}\beta_{K-N+2,N-1}$; $H_1 : (\tilde{t}_{\mathrm{K}} \text{ given } \beta) \sim \mathcal{CF}_{1,K-N+1}(\delta)$, $\delta^2 = \mathrm{SNR}\,\beta\,\cos^2\theta$, $\beta \sim \mathcal{C}\beta_{K-N+2,N-1}(\delta_\beta)$, $\delta_\beta^2 = \mathrm{SNR}\,\sin^2\theta$;
AMF (t_{AMF})	$H_0 : t_{\mathrm{AMF}} = \dfrac{\tilde{t}_{\mathrm{K}}}{\beta} \sim \dfrac{\mathcal{CF}_{1,K-N+1}}{\mathcal{C}\beta_{K-N+2,N-1}}$; $H_1 : (t_{\mathrm{AMF}} \text{ given } \beta) \sim \dfrac{\mathcal{CF}_{1,K-N+1}(\delta)}{\beta}$;
ACE $\left(\tilde{t}_{\mathrm{ACE}} = \dfrac{t_{\mathrm{ACE}}}{1 - t_{\mathrm{ACE}}}\right)$	$H_0 : \tilde{t}_{\mathrm{ACE}} = \dfrac{\tilde{t}_{\mathrm{K}}}{1-\beta} \sim \dfrac{\mathcal{CF}_{1,K-N+1}}{1-\mathcal{C}\beta_{K-N+2,N-1}}$; $H_1 : (\tilde{t}_{\mathrm{ACE}} \text{ given } \beta) \sim \dfrac{\mathcal{CF}_{1,K-N+1}(\delta)}{1-\beta}$;
W-ABORT (t_{WA})	$H_0 : t_{\mathrm{WA}} = (1+\tilde{t}_{\mathrm{K}})\beta \sim \left(1+\mathcal{CF}_{1,K-N+1}\right)\mathcal{C}\beta_{K-N+2,N-1}$; $H_1 : (t_{\mathrm{WA}} \text{ given } \beta) \sim \left(1+\mathcal{CF}_{1,K-N+1}(\delta)\right)\beta$.

APPENDIX C

Statistical Characterization of S-ASB and WAS-ASB

This appendix contains the statistical characterization of t_{SD} and β which allows to characterize the two-stage detectors presented in Chapter 4. As a first step, suitable stochastic representations for t_{SD} and β are derived.

C.1 STOCHASTIC REPRESENTATIONS FOR t_{SD} AND β

First, define the following equivalent statistic for the SD (3.13)

$$\tilde{t}_{\mathrm{SD}} = \frac{1}{1-t_{\mathrm{SD}}} = \frac{1+\boldsymbol{r}^{\dagger}\boldsymbol{S}^{-1}\boldsymbol{r}}{1+\boldsymbol{r}^{\dagger}\boldsymbol{S}^{-1}\boldsymbol{r}-\boldsymbol{r}^{\dagger}\boldsymbol{S}^{-1}\boldsymbol{H}(\boldsymbol{H}^{\dagger}\boldsymbol{S}^{-1}\boldsymbol{H})^{-1}\boldsymbol{H}^{\dagger}\boldsymbol{S}^{-1}\boldsymbol{r}}.$$

Then, rewrite $\tilde{t}_{\mathrm{SD}}$, β, and t_{K} in terms of the whitened quantities $\boldsymbol{r}_w = \boldsymbol{M}^{-1/2}\boldsymbol{r}$, $\boldsymbol{v}_w = \boldsymbol{M}^{-1/2}\boldsymbol{v}$, $\boldsymbol{H}_w = \boldsymbol{M}^{-1/2}\boldsymbol{H}$, and $\boldsymbol{S}_w = \boldsymbol{M}^{-1/2}\boldsymbol{S}\boldsymbol{M}^{-1/2}$, as

$$\tilde{t}_{\mathrm{SD}} = \frac{1+\boldsymbol{r}_w^{\dagger}\boldsymbol{S}_w^{-1}\boldsymbol{r}_w}{1+\boldsymbol{r}_w^{\dagger}\boldsymbol{S}_w^{-1}\boldsymbol{r}_w-\boldsymbol{r}_w^{\dagger}\boldsymbol{S}_w^{-1}\boldsymbol{H}_w(\boldsymbol{H}_w^{\dagger}\boldsymbol{S}_w^{-1}\boldsymbol{H}_w)^{-1}\boldsymbol{H}_w^{\dagger}\boldsymbol{S}_w^{-1}\boldsymbol{r}_w},$$

$$\beta = \frac{1}{1+\boldsymbol{r}_w^{\dagger}\boldsymbol{S}_w^{-1}\boldsymbol{r}_w - \dfrac{|\boldsymbol{r}_w^{\dagger}\boldsymbol{S}_w^{-1}\boldsymbol{v}_w|^2}{\boldsymbol{v}_w^{\dagger}\boldsymbol{S}_w^{-1}\boldsymbol{v}_w}},$$

and

$$t_{\mathrm{K}} = \frac{|\boldsymbol{r}_w^{\dagger}\boldsymbol{S}_w^{-1}\boldsymbol{v}_w|^2}{(1+\boldsymbol{r}_w^{\dagger}\boldsymbol{S}_w^{-1}\boldsymbol{r}_w)(\boldsymbol{v}_w^{\dagger}\boldsymbol{S}_w^{-1}\boldsymbol{v}_w)},$$

respectively.

Since the nominal steering vector $\boldsymbol{v}$ is assumed to belong to $\langle \boldsymbol{H} \rangle$, without loss of generality we can choose the matrix $\boldsymbol{H}$ such that its first column is $\boldsymbol{v}$. Thus, after re-writing $\boldsymbol{H}_w$ in terms of its QR factorization, i.e.,

$$\boldsymbol{H}_w = \boldsymbol{H}_0\boldsymbol{T}_H,$$

with $\boldsymbol{H}_0 \in \mathbb{C}^{N\times r}$ a slice of unitary matrix, namely $\boldsymbol{H}_0^{\dagger}\boldsymbol{H}_0 = \boldsymbol{I}$, and $\boldsymbol{T}_H \in \mathbb{C}^{r\times r}$ an invertible upper triangular matrix, we can define a unitary matrix[1] $\boldsymbol{U} \in \mathbb{C}^{N\times N}$ that rotates the r orthonormal columns of $\boldsymbol{H}_0$ onto the first r vectors of the standard basis of $\mathbb{C}^{N\times 1}$, i.e.,

$$\boldsymbol{U}\boldsymbol{H}_0 = \boldsymbol{E}_r = [\boldsymbol{e}_1 \cdots \boldsymbol{e}_r] = \begin{bmatrix} \boldsymbol{I} \\ \boldsymbol{0} \end{bmatrix}$$

[1] $\boldsymbol{U}$ is the same matrix introduced in Section 3.2.

and

$$\boldsymbol{U}\boldsymbol{v}_w = \boldsymbol{U}\boldsymbol{M}^{-1/2}\boldsymbol{v} = \sqrt{\boldsymbol{v}^\dagger \boldsymbol{M}^{-1}\boldsymbol{v}}\, \boldsymbol{e}_1.$$

Accordingly, the test statistics can be recast as follows

$$\begin{aligned}
\tilde{t}_{\mathrm{SD}} &= \frac{1+\boldsymbol{x}^\dagger \boldsymbol{S}_1^{-1}\boldsymbol{x}}{1+\boldsymbol{x}^\dagger \boldsymbol{S}_1^{-1}\boldsymbol{x} - \boldsymbol{x}^\dagger \boldsymbol{S}_1^{-1}\boldsymbol{E}_r(\boldsymbol{E}_r^\dagger \boldsymbol{S}_1^{-1}\boldsymbol{E}_r)^{-1}\boldsymbol{E}_r^\dagger \boldsymbol{S}_1^{-1}\boldsymbol{x}}, \\
\beta &= \frac{1}{1+\boldsymbol{x}^\dagger \boldsymbol{S}_1^{-1}\boldsymbol{x} - \dfrac{|\boldsymbol{x}^\dagger \boldsymbol{S}_1^{-1}\boldsymbol{e}_1|^2}{\boldsymbol{e}_1^\dagger \boldsymbol{S}_1^{-1}\boldsymbol{e}_1}}, \\
t_{\mathrm{K}} &= \frac{|\boldsymbol{x}^\dagger \boldsymbol{S}_1^{-1}\boldsymbol{e}_1|^2}{(1+\boldsymbol{x}^\dagger \boldsymbol{S}_1^{-1}\boldsymbol{x})(\boldsymbol{e}_1^\dagger \boldsymbol{S}_1^{-1}\boldsymbol{e}_1)},
\end{aligned} \tag{C.1}$$

where $\boldsymbol{x} = \boldsymbol{U}\boldsymbol{r}_w \in \mathbb{C}^{N\times 1}$ and $\boldsymbol{S}_1 = \boldsymbol{U}\boldsymbol{S}_w\boldsymbol{U}^\dagger \in \mathbb{C}^{N\times N}$.

Moreover, the term $\boldsymbol{x}^\dagger \boldsymbol{S}_1^{-1}\boldsymbol{x}$ can be rewritten in two different forms. To this end, let $\boldsymbol{Q} \in \mathbb{C}^{m\times m}$ and $\boldsymbol{q} \in \mathbb{C}^{m\times 1}$. For index sets $A, B \subseteq \{1,\ldots,m\}$, we denote the (sub)matrix that lies in the rows of $\boldsymbol{Q}$ indexed by A and the columns indexed by B as $\boldsymbol{Q}_{AB}$; analogously, we denote the (sub)vector that lies in the rows of $\boldsymbol{q}$ indexed by A as $\boldsymbol{q}_A$ [53].

Now let

$$A = \{1\}, \quad B = \{2,\cdots,r\}, \quad \text{and} \quad C = \{r+1,\cdots,N\}.$$

It follows that the (Hermitian) matrix $\boldsymbol{S}_1^{-1}$ and the vector $\boldsymbol{x}$ can be rewritten as

$$\boldsymbol{S}_1^{-1} = \begin{bmatrix} \underset{1,1}{C_{AA}} & \underset{1,r-1}{\boldsymbol{C}_{AB}} & \underset{1,N-r}{\boldsymbol{C}_{AC}} \\ \underset{r-1,1}{\boldsymbol{C}_{BA}} & \underset{r-1,r-1}{\boldsymbol{C}_{BB}} & \underset{r-1,N-r}{\boldsymbol{C}_{BC}} \\ \underset{N-r,1}{\boldsymbol{C}_{CA}} & \underset{N-r,r-1}{\boldsymbol{C}_{CB}} & \underset{N-r,N-r}{\boldsymbol{C}_{CC}} \end{bmatrix} \quad \text{and} \quad \boldsymbol{x} = \begin{bmatrix} \underset{1,1}{x_A} \\ \underset{r-1,1}{\boldsymbol{x}_B} \\ \underset{N-r,1}{\boldsymbol{x}_C} \end{bmatrix},$$

respectively. Observe also that

$$\boldsymbol{x} = \begin{bmatrix} x_A \\ \boldsymbol{x}_{B\cup C} \end{bmatrix} \quad \text{with} \quad \boldsymbol{x}_{B\cup C} = \begin{bmatrix} \boldsymbol{x}_B \\ \boldsymbol{x}_C \end{bmatrix} \in \mathbb{C}^{(N-1)\times 1}.$$

Let also

$$\boldsymbol{S}_1^{-1} = \begin{bmatrix} A_1 & \boldsymbol{B}_1 \\ \boldsymbol{C}_1 & \boldsymbol{D}_1 \end{bmatrix},$$

where

$$\begin{aligned}
A_1 &= C_{AA} \in \mathbb{C}, \ \boldsymbol{B}_1 = [\boldsymbol{C}_{AB}\ \boldsymbol{C}_{AC}] \in \mathbb{C}^{1\times(N-1)}, \\
\boldsymbol{C}_1 &= \begin{bmatrix} \boldsymbol{C}_{BA} \\ \boldsymbol{C}_{CA} \end{bmatrix} = \boldsymbol{B}_1^\dagger \in \mathbb{C}^{(N-1)\times 1},
\end{aligned}$$

and

$$D_1 = \begin{bmatrix} C_{BB} & C_{BC} \\ C_{CB} & C_{CC} \end{bmatrix} \in \mathbb{C}^{(N-1)\times(N-1)}.$$

Using the above partitioned forms of S_1^{-1} and x to evaluate $x^\dagger S_1^{-1} x$, we get

$$\begin{aligned} x^\dagger S_1^{-1} x &= x_A^* A_1 x_A + 2\Re e\{x_A^* B_1 x_{B\cup C}\} + x_{B\cup C}^\dagger B_1^\dagger A_1^{-1} B_1 x_{B\cup C} \\ &\quad + x_{B\cup C}^\dagger D_1 x_{B\cup C} - x_{B\cup C}^\dagger B_1^\dagger A_1^{-1} B_1 x_{B\cup C} \\ &= (A_1 x_A + B_1 x_{B\cup C})^* A_1^{-1} (A_1 x_A + B_1 x_{B\cup C}) \\ &\quad + x_{B\cup C}^\dagger (D_1 - B_1^\dagger A_1^{-1} B_1) x_{B\cup C} \\ &= (C_{AA} x_A + C_{AB} x_B + C_{AC} x_C)^* C_{AA}^{-1} (C_{AA} x_A + C_{AB} x_B + C_{AC} x_C) \\ &\quad + x_{B\cup C}^\dagger (D_1 - B_1^\dagger A_1^{-1} B_1) x_{B\cup C}. \end{aligned} \tag{C.2}$$

At this point it is possible to recast the statistic of Kelly's detector as follows:

$$\begin{aligned} t_K = {} & \frac{|x_A^* C_{AA} + x_B^\dagger C_{BA} + x_C^\dagger C_{CA}|^2}{C_{AA}} \\ & \times \left[1 + \frac{|x_A^* C_{AA} + x_B^\dagger C_{BA} + x_C^\dagger C_{CA}|^2}{C_{AA}} + x_{B\cup C}^\dagger (D_1 - B_1^\dagger A_1^{-1} B_1) x_{B\cup C} \right]^{-1}, \end{aligned} \tag{C.3}$$

where use has been made of (C.2) and of the following identities:

$$e_1^\dagger S_1^{-1} e_1 = C_{AA}, \tag{C.4}$$

$$x^\dagger S_1^{-1} e_1 = (C_{AA} x_A + C_{AB} x_B + C_{AC} x_C)^*. \tag{C.5}$$

An alternative representation for $x^\dagger S_1^{-1} x$ can be obtained by partitioning x and S_1^{-1} as follows:

$$x = \begin{bmatrix} x_{A\cup B} \\ x_C \end{bmatrix} \quad \text{and} \quad S_1^{-1} = \begin{bmatrix} A_2 & B_2 \\ C_2 & D_2 \end{bmatrix},$$

where

$$A_2 = A_2^\dagger = \begin{bmatrix} C_{AA} & C_{AB} \\ C_{BA} & C_{BB} \end{bmatrix} \in \mathbb{C}^{r\times r}, \; B_2 = \begin{bmatrix} C_{AC} \\ C_{BC} \end{bmatrix} \in \mathbb{C}^{r\times(N-r)},$$

$$C_2 = [C_{CA}\; C_{CB}] = B_2^\dagger \in \mathbb{C}^{(N-r)\times r}, \text{ and } D_2 = C_{CC} \in \mathbb{C}^{(N-r)\times(N-r)}.$$

Thus, we obtain

$$\begin{aligned} x^\dagger S_1^{-1} x &= x_{A\cup B}^\dagger A_2 x_{A\cup B} + 2\Re e\{x_C^\dagger B_2^\dagger x_{A\cup B}\} + x_C^\dagger B_2^\dagger A_2^{-1} B_2 x_C \\ &\quad + x_C^\dagger D_2 x_C - x_C^\dagger B_2^\dagger A_2^{-1} B_2 x_C \\ &= (A_2 x_{A\cup B} + B_2 x_C)^\dagger A_2^{-1} (A_2 x_{A\cup B} + B_2 x_C) \\ &\quad + x_C^\dagger (D_2 - B_2^\dagger A_2^{-1} B_2) x_C, \end{aligned} \tag{C.6}$$

$$E_r^\dagger S_1^{-1} x = A_2 x_{A\cup B} + B_2 x_C, \tag{C.7}$$

$$E_r^\dagger S_1^{-1} E_r = A_2. \tag{C.8}$$

Inserting (C.2) into the numerator of (C.1) and (C.6), (C.7), and (C.8) into the denominator of (C.1), yields

$$\tilde{t}_{\mathrm{SD}} = \frac{1+\boldsymbol{x}_{B\cup C}^{\dagger}(\boldsymbol{D}_1-\boldsymbol{B}_1^{\dagger}A_1^{-1}\boldsymbol{B}_1)\boldsymbol{x}_{B\cup C}}{1+\boldsymbol{x}_C^{\dagger}(\boldsymbol{D}_2-\boldsymbol{B}_2^{\dagger}A_2^{-1}\boldsymbol{B}_2)\boldsymbol{x}_C}\left[\frac{|x_A^{*}C_{AA}+\boldsymbol{x}_B^{\dagger}\boldsymbol{C}_{BA}+\boldsymbol{x}_C^{\dagger}\boldsymbol{C}_{CA}|^2/C_{AA}}{1+\boldsymbol{x}_{B\cup C}^{\dagger}(\boldsymbol{D}_1-\boldsymbol{B}_1^{\dagger}A_1^{-1}\boldsymbol{B}_1)\boldsymbol{x}_{B\cup C}}+1\right]. \quad \text{(C.9)}$$

Remember now that

$$\boldsymbol{S}_1 = \begin{bmatrix} A_1 & \boldsymbol{B}_1 \\ \boldsymbol{C}_1 & \boldsymbol{D}_1 \end{bmatrix}^{-1} = \begin{bmatrix} A_2 & \boldsymbol{B}_2 \\ \boldsymbol{C}_2 & \boldsymbol{D}_2 \end{bmatrix}^{-1}.$$

Thus, resorting to the formulas for the inverse of a partitioned matrix [52] yields

$$(\boldsymbol{D}_1-\boldsymbol{B}_1^{\dagger}A_1^{-1}\boldsymbol{B}_1) = \begin{bmatrix} \boldsymbol{S}_{1_{BB}} & \boldsymbol{S}_{1_{BC}} \\ \boldsymbol{S}_{1_{CB}} & \boldsymbol{S}_{1_{CC}} \end{bmatrix}^{-1}, \qquad (\boldsymbol{D}_2-\boldsymbol{B}_2^{\dagger}A_2^{-1}\boldsymbol{B}_2) = \boldsymbol{S}_{1_{CC}}^{-1};$$

as a consequence, we have that

$$\boldsymbol{x}_C^{\dagger}(\boldsymbol{D}_2-\boldsymbol{B}_2^{\dagger}A_2^{-1}\boldsymbol{B}_2)\boldsymbol{x}_C = \boldsymbol{x}_C^{\dagger}\boldsymbol{S}_{1_{CC}}^{-1}\boldsymbol{x}_C \quad \text{(C.10)}$$

and

$$\begin{aligned}\boldsymbol{x}_{B\cup C}^{\dagger}(\boldsymbol{D}_1-\boldsymbol{B}_1^{\dagger}A_1^{-1}\boldsymbol{B}_1)\boldsymbol{x}_{B\cup C} &= \boldsymbol{x}_{B\cup C}^{\dagger}\begin{bmatrix} \boldsymbol{S}_{1_{BB}} & \boldsymbol{S}_{1_{BC}} \\ \boldsymbol{S}_{1_{CB}} & \boldsymbol{S}_{1_{CC}} \end{bmatrix}^{-1}\boldsymbol{x}_{B\cup C} \\ &= (\boldsymbol{x}_B-\boldsymbol{S}_{1_{BC}}\boldsymbol{S}_{1_{CC}}^{-1}\boldsymbol{x}_C)^{\dagger}(\boldsymbol{S}_{1_{BB}}-\boldsymbol{S}_{1_{BC}}\boldsymbol{S}_{1_{CC}}^{-1}\boldsymbol{S}_{1_{CB}})^{-1}(\boldsymbol{x}_B-\boldsymbol{S}_{1_{BC}}\boldsymbol{S}_{1_{CC}}^{-1}\boldsymbol{x}_C) \\ &\quad + \boldsymbol{x}_C^{\dagger}\boldsymbol{S}_{1_{CC}}^{-1}\boldsymbol{x}_C, \end{aligned} \quad \text{(C.11)}$$

where we have also used formula (2) of paragraph (3.5.3) in [52].

Substituting equations (C.10) and (C.11) into (C.9) and (C.11) into (C.3), after some algebra, yields

$$\tilde{t}_{\mathrm{SD}} = (1+c)\left(\tilde{t}_{\mathrm{K}}+1\right),$$

where

$$c = \frac{(\boldsymbol{x}_B-\boldsymbol{S}_{1_{BC}}\boldsymbol{S}_{1_{CC}}^{-1}\boldsymbol{x}_C)^{\dagger}(\boldsymbol{S}_{1_{BB}}-\boldsymbol{S}_{1_{BC}}\boldsymbol{S}_{1_{CC}}^{-1}\boldsymbol{S}_{1_{CB}})^{-1}(\boldsymbol{x}_B-\boldsymbol{S}_{1_{BC}}\boldsymbol{S}_{1_{CC}}^{-1}\boldsymbol{x}_C)}{1+b}, \quad \text{(C.12)}$$

with

$$b = \boldsymbol{x}_C^{\dagger}\boldsymbol{S}_{1_{CC}}^{-1}\boldsymbol{x}_C \quad \text{(C.13)}$$

and

$$\begin{aligned}\tilde{t}_{\mathrm{K}} &= \frac{|x_A^{*}C_{AA}+\boldsymbol{x}_B^{\dagger}\boldsymbol{C}_{BA}+\boldsymbol{x}_C^{\dagger}\boldsymbol{C}_{CA}|^2/C_{AA}}{\left[1+(\boldsymbol{x}_B-\boldsymbol{S}_{1_{BC}}\boldsymbol{S}_{1_{CC}}^{-1}\boldsymbol{x}_C)^{\dagger}(\boldsymbol{S}_{1_{BB}}-\boldsymbol{S}_{1_{BC}}\boldsymbol{S}_{1_{CC}}^{-1}\boldsymbol{S}_{1_{CB}})^{-1}(\boldsymbol{x}_B-\boldsymbol{S}_{1_{BC}}\boldsymbol{S}_{1_{CC}}^{-1}\boldsymbol{x}_C)+b\right]} \\ &= \frac{t_{\mathrm{K}}}{1-t_{\mathrm{K}}}.\end{aligned}$$

As to β, we can resort to (C.2), (C.4), (C.5), and (C.11) to recast it as

$$\begin{aligned}\beta &= \left[1 + \boldsymbol{x}_{B\cup C}^{\dagger}(\boldsymbol{D}_1 - \boldsymbol{B}_1^{\dagger}A_1^{-1}\boldsymbol{B}_1)\boldsymbol{x}_{B\cup C}\right]^{-1} \\ &= \left[1 + (\boldsymbol{x}_B - \boldsymbol{S}_{1_{BC}}\boldsymbol{S}_{1_{CC}}^{-1}\boldsymbol{x}_C)^{\dagger}(\boldsymbol{S}_{1_{BB}} - \boldsymbol{S}_{1_{BC}}\boldsymbol{S}_{1_{CC}}^{-1}\boldsymbol{S}_{1_{CB}})^{-1} \right. \\ &\quad \left. \times (\boldsymbol{x}_B - \boldsymbol{S}_{1_{BC}}\boldsymbol{S}_{1_{CC}}^{-1}\boldsymbol{x}_C) + \boldsymbol{x}_C^{\dagger}\boldsymbol{S}_{1_{CC}}^{-1}\boldsymbol{x}_C\right]^{-1} \\ &= \frac{1}{(1+b)(1+c)}. \end{aligned} \tag{C.14}$$

In order to show that the expression of $\tilde{t}_{\text{SD}}$ in terms of $\tilde{t}_{\text{K}}$ and c (and eventually b) can be used to evaluate the performance of S-ASB and WAS-ASB, it is sufficient to observe the following two facts

- both the ACE and the W-ABORT can be expressed in terms of $\tilde{t}_{\text{K}}$ and $\beta = \frac{1}{(1+b)(1+c)}$ (see Appendix B);
- the statistical characterization of $\tilde{t}_{\text{K}}$ conditioned on β (see Appendix B) is also the statistical characterization of $\tilde{t}_{\text{K}}$ conditioned on b and c; in fact, the distribution of $\tilde{t}_{\text{K}}$ conditioned on the B and C components of the vectors $\boldsymbol{x} = \boldsymbol{U}\boldsymbol{M}^{-1/2}\boldsymbol{r}$ and $\boldsymbol{x}_k = \boldsymbol{U}\boldsymbol{M}^{-1/2}\boldsymbol{r}_k, k = 1, \ldots, K$, depends on β only [2, 37, 38].

Now it only remains to jointly characterize b and c under both hypotheses. This is the object of the next section.

C.2 STATISTICAL CHARACTERIZATION OF b AND c

In the following we derive the statistical characterization (under H_0 and H_1) of the random variables b and c, given by equations (C.13) and (C.12), respectively.

C.2.1 DISTRIBUTIONS UNDER H_0

Let us assume that H_0 is in force, namely

$$\boldsymbol{r} = \boldsymbol{n}, \quad \text{with} \quad \boldsymbol{n} \sim \mathcal{CN}_N(\boldsymbol{0}, \boldsymbol{M}).$$

The whitening transformation $\boldsymbol{M}^{-1/2}$ and the rotation $\boldsymbol{U}$ applied to vectors $\boldsymbol{r}$ and $\boldsymbol{r}_k, k = 1, \ldots, K$, lead to

$$\begin{aligned}\boldsymbol{x} &= \boldsymbol{U}\boldsymbol{M}^{-1/2}\boldsymbol{r} \sim \mathcal{CN}_N(\boldsymbol{0}, \boldsymbol{I}), \\ \boldsymbol{S}_1 &= \boldsymbol{U}\boldsymbol{M}^{-1/2}\boldsymbol{S}\boldsymbol{M}^{-1/2}\boldsymbol{U}^{\dagger} \sim \mathcal{CW}_N(K, \boldsymbol{I}).\end{aligned}$$

The statistical characterization of b descends directly from Theorem A.13, as it is proved below.

Proposition C.1. ***Statistical characterization of b under H_0*** *Under the H_0 hypothesis, the rv b, given by* (C.13), *is a complex central F-distribution with $N - r$, $K - N + r + 1$ degrees of freedom.*

Proof. First notice that, since $\boldsymbol{x}$ and $\boldsymbol{S}_1$ are statistically independent, also $\boldsymbol{x}_C$ and $\boldsymbol{S}_{1_{CC}}$ are independent random quantities; in addition, $\boldsymbol{x}_C \sim \mathcal{CN}_{N-r}(\boldsymbol{0}, \boldsymbol{I})$ and, by Theorem A.11, $\boldsymbol{S}_{1_{CC}} \sim \mathcal{CW}_{N-r}(K, \boldsymbol{I})$. Now, define the following random variable

$$q = \frac{\boldsymbol{x}_C^\dagger \boldsymbol{x}_C}{\boldsymbol{x}_C^\dagger \boldsymbol{S}_{1_{CC}}^{-1} \boldsymbol{x}_C}. \tag{C.15}$$

By Theorem A.13, it follows that

- q is statistically independent of $\boldsymbol{x}_C$;
- q is a complex central chi-square with $K-(N-r)+1$ degrees of freedom, i.e., $q \sim \mathcal{C}\chi^2_{K-N+r+1}$.

Finally, b can be written as the ratio between two independent complex central chi-square random variables, namely as

$$b = \boldsymbol{x}_C^\dagger \boldsymbol{S}_{1_{CC}}^{-1} \boldsymbol{x}_C = \frac{\boldsymbol{x}_C^\dagger \boldsymbol{x}_C}{(\boldsymbol{x}_C^\dagger \boldsymbol{x}_C)/(\boldsymbol{x}_C^\dagger \boldsymbol{S}_{1_{CC}}^{-1} \boldsymbol{x}_C)} = \frac{\boldsymbol{x}_C^\dagger \boldsymbol{x}_C}{q},$$

where $\boldsymbol{x}_C^\dagger \boldsymbol{x}_C \sim \mathcal{C}\chi^2_{N-r}$. Thus, by Proposition A.6, we conclude that $b \sim \mathcal{CF}_{N-r,K-N+r+1}$. □

As to the distribution of c under H_0, the following proposition holds true.

Proposition C.2. ***Statistical characterization of*** c ***under*** H_0 *Under the* H_0 *hypothesis, the rv* c*, given by* (C.12)*, is a complex central F-distribution with* $r-1$, $K-N+2$ *degrees of freedom and is independent of* b.

Proof. First, following the lead of [2], we recast c as

$$c = \frac{\boldsymbol{y}^\dagger \boldsymbol{G}^{-1} \boldsymbol{y}}{1+b},$$

where

$$\boldsymbol{y} = \boldsymbol{x}_B - \boldsymbol{S}_{1_{BC}} \boldsymbol{S}_{1_{CC}}^{-1} \boldsymbol{x}_C \in \mathbb{C}^{(r-1)\times 1},$$
$$\boldsymbol{G} = \boldsymbol{S}_{1_{BB}} - \boldsymbol{S}_{1_{BC}} \boldsymbol{S}_{1_{CC}}^{-1} \boldsymbol{S}_{1_{CB}} \in \mathbb{C}^{(r-1)\times(r-1)}.$$

Second, recalling that $\boldsymbol{x}_k = \boldsymbol{U}\boldsymbol{M}^{-1/2}\boldsymbol{r}_k$, $k=1,\ldots,K$, it follows that

$$\boldsymbol{x}_{k_{B\cup C}} = (\boldsymbol{U}\boldsymbol{M}^{-1/2}\boldsymbol{r}_k)_{B\cup C} \sim \mathcal{CN}_{N-1}(\boldsymbol{0}, \boldsymbol{I}), \quad k=1,\ldots,K,$$

and that such vectors are independent; thus, by Theorem A.11, it follows that

$$\boldsymbol{G} \sim \mathcal{CW}_{r-1}(K-(N-1)+(r-1), \boldsymbol{I})$$

and it is independent of $\{S_{1_{BC}}, S_{1_{CC}}\}$. Obviously, $\{G, S_{1_{BC}}, S_{1_{CC}}\}$ is also independent of the data under test and, hence, of $\{x_B, x_C\}$. Gathering these results together, it follows that G is independent of $\{S_{1_{BC}}, S_{1_{CC}}, x_B, x_C\}$ and, hence, of $\{v, b\}$, where

$$v = \frac{y}{\sqrt{1+b}}.$$

Third, it is possible to show that, given the C components,

$$y \sim \mathcal{CN}_{r-1}(\mathbf{0}, (1+b)I). \tag{C.16}$$

Leaving aside for the moment the proof of (C.16), given the C components, it follows that

$$v = \frac{y}{\sqrt{1+b}} \sim \mathcal{CN}_{r-1}(\mathbf{0}, I)$$

and, since the conditional distribution of v is independent of the C components, it is also the unconditional one; as a by-product v is independent of b. It follows that the random quantities G, v, and b are independent.

We can now conclude that $c = v^\dagger G^{-1} v$ is independent of b and, by Theorem A.13, that

$$c = v^\dagger G^{-1} v = \frac{v^\dagger v}{\dfrac{v^\dagger v}{v^\dagger G^{-1} v}} \sim \mathcal{CF}_{r-1, K-N+2}.$$

It still remains to prove that, given the C components,

$$y \sim \mathcal{CN}_{r-1}(\mathbf{0}, (1+b)I);$$

to this end, observe that, given the C components, y is a linear combination of independent complex normal random vectors with zero mean and covariance matrix I; in fact, it is given by

$$y = x_B - \sum_{i=1}^{K} x_{iB} x_{iC}^\dagger S_{1_{CC}}^{-1} x_C.$$

Moreover, denoting by $E_C[\cdot]$ statistical expectation conditioned on the C components, the mean vector and the covariance matrix of y are given by

$$\begin{aligned}
E_C[y] &= \mathbf{0},\\
E_C[y y^\dagger] &= I + I \sum_{i=1}^{K} x_{iC}^\dagger S_{1_{CC}}^{-1} x_C x_C^\dagger S_{1_{CC}}^{-1} x_{iC}\\
&= I + I\left[x_C^\dagger S_{1_{CC}}^{-1} \sum_{i=1}^{K} x_{iC} x_{iC}^\dagger S_{1_{CC}}^{-1} x_C\right]\\
&= I + I x_C^\dagger S_{1_{CC}}^{-1} x_C\\
&= (1+b)I.
\end{aligned}$$

An alternative proof of the last equation can be found in [2]. □

C.2.2 DISTRIBUTIONS UNDER H_1

Suppose that the H_1 hypothesis (given by equation (3.1)) is in force, i.e., the actual steering vector $\boldsymbol{p}$ and the nominal one $\boldsymbol{v}$ are not necessarily aligned:

$$\boldsymbol{r} = \alpha \boldsymbol{p} + \boldsymbol{n}, \quad \text{with} \quad \boldsymbol{n} \sim \mathcal{CN}_N(\boldsymbol{0}, \boldsymbol{M}).$$

Then, using equations (3.3) and (3.14), the random vector $\boldsymbol{x} = \boldsymbol{U}\boldsymbol{M}^{-1/2}\boldsymbol{r}$ can be written as

$$\boldsymbol{x} = \alpha \boldsymbol{U}\boldsymbol{M}^{-1/2}\boldsymbol{p} + \boldsymbol{U}\boldsymbol{M}^{-1/2}\boldsymbol{n} = \alpha\sqrt{\boldsymbol{p}^\dagger \boldsymbol{M}^{-1}\boldsymbol{p}} \begin{bmatrix} e^{j\xi}\cos\theta \\ \boldsymbol{h}_B \sin\theta \\ \boldsymbol{h}_C \sin\theta \end{bmatrix} + \boldsymbol{U}\boldsymbol{M}^{-1/2}\boldsymbol{n},$$

where $\boldsymbol{h}_B \in \mathbb{C}^{(r-1)\times 1}$ and $\boldsymbol{h}_C \in \mathbb{C}^{(N-r)\times 1}$ are such that $\|\boldsymbol{h}_B\|^2 + \|\boldsymbol{h}_C\|^2 = 1$ and $e^{j\xi}\cos\theta$ is defined by equation (3.6). It is now straightforward to show that $\boldsymbol{x}$ is distributed as

$$\boldsymbol{x} \sim \mathcal{CN}_N\left(\alpha\sqrt{\boldsymbol{p}^\dagger \boldsymbol{M}^{-1}\boldsymbol{p}} \begin{bmatrix} e^{j\xi}\cos\theta \\ \boldsymbol{h}_B \sin\theta \\ \boldsymbol{h}_C \sin\theta \end{bmatrix}, \boldsymbol{I}\right).$$

Since the rv's b and c depend on the mismatch angle θ we will denote these rv's by b_θ and c_θ.

Proposition C.3. ***Statistical characterization of*** b_θ ***under*** H_1 *Under the H_1 hypothesis, the rv b_θ is a complex noncentral F-distribution with $N-r$, $K-N+r+1$ degrees of freedom and noncentrality parameter δ_{b_θ}, with*

$$\delta^2_{b_\theta} = \text{SNR}\, \|\boldsymbol{h}_C\|^2 \sin^2\theta,$$

where the SNR *is given by* (B.2).

Proof. Rewrite b_θ as follows

$$b_\theta = \frac{\boldsymbol{x}_C^\dagger \boldsymbol{x}_C}{q},$$

where q is given by (C.15); now observe that

$$\begin{aligned} \boldsymbol{x}_C &\sim \mathcal{CN}_{N-r}(\alpha\sqrt{\boldsymbol{p}^\dagger \boldsymbol{M}^{-1}\boldsymbol{p}}\, \boldsymbol{h}_C \sin\theta, \boldsymbol{I}), \\ \boldsymbol{S}_{1_{CC}} &\sim \mathcal{CW}_{N-r}(K, \boldsymbol{I}), \end{aligned}$$

and $\boldsymbol{x}_c$ and $\boldsymbol{S}_{1_{CC}}$ are independent. Thus, by Theorem A.13 and Proposition A.6 it follows that the random variable b_θ is ruled by the complex noncentral F-distribution with $N-r$, $K-N+r+1$ degrees of freedom and noncentrality parameter δ_{b_θ}. □

Proposition C.4. ***Statistical characterization of c_θ under H_1*** *Under the H_1 hypothesis and given b_θ, the rv c_θ is ruled by the complex noncentral F-distribution with $r-1$, $K-N+2$ degrees of freedom and noncentrality parameter δ_{c_θ}, with*

$$\delta^2_{c_\theta} = \frac{\text{SNR}\,\|\boldsymbol{h}_B\|^2\ \sin^2\theta}{1+b_\theta},$$

where the SNR *is defined by equation* (B.2).

Proof. Let us begin recasting c_θ as follows

$$c_\theta = \boldsymbol{v}^\dagger \boldsymbol{G}^{-1}\boldsymbol{v},$$

where

$$\boldsymbol{v} = \frac{\boldsymbol{x}_B - \boldsymbol{S}_{1_{BC}}\boldsymbol{S}^{-1}_{1_{CC}}\boldsymbol{x}_C}{\sqrt{1+b_\theta}},$$

and

$$\boldsymbol{G} = \boldsymbol{S}_{1_{BB}} - \boldsymbol{S}_{1_{BC}}\boldsymbol{S}^{-1}_{1_{CC}}\boldsymbol{S}_{1_{CB}};$$

following the rationale of the proof of Proposition C.2, it can be easily shown that

- $\boldsymbol{G} \sim \mathcal{CW}_{r-1}(K-N+r, \boldsymbol{I})$; moreover it is independent of $\{\boldsymbol{S}_{1_{BC}}, \boldsymbol{S}_{1_{CC}}, \boldsymbol{x}_B, \boldsymbol{x}_C\}$ and, hence, of $\{\boldsymbol{v}, b_\theta\}$; as a consequence, given b_θ, $\boldsymbol{G}$ and $\boldsymbol{v}$ are independent.
- given the C-components

  $$\boldsymbol{v} \sim \mathcal{CN}_{r-1}\left(\alpha\sqrt{\frac{\boldsymbol{p}^\dagger \boldsymbol{M}^{-1}\boldsymbol{p}}{1+b_\theta}}\boldsymbol{h}_B \sin\theta, \boldsymbol{I}\right),$$

  that is also the characterization of $\boldsymbol{v}$ given b_θ.

Therefore, by Theorem A.13 and Proposition A.6 we have that

$$c_\theta = \frac{\boldsymbol{v}^\dagger\boldsymbol{v}}{\boldsymbol{v}^\dagger\boldsymbol{v}/\boldsymbol{v}^\dagger\boldsymbol{G}^{-1}\boldsymbol{v}},$$

given b_θ, is ruled by the complex noncentral F-distribution with $r-1$, $K-N+2$ degrees of freedom and noncentrality parameter δ_{c_θ}. □

In Table C.1 we summarize the distributions of the random variables b and c under both hypotheses.

Table C.1: Distributions of b and c.

Random Variable	Statistical characterization
b (or b_θ under H_1)	$H_0 : \mathcal{CF}_{N-r,K-N+r+1}$; $H_1 : \mathcal{CF}_{N-r,K-N+r+1}(\delta_{b_\theta})$, $\delta^2_{b_\theta} = \text{SNR}\ \lVert \boldsymbol{h}_C \rVert^2\ \sin^2\theta$;
c (or c_θ under H_1)	$H_0 : \mathcal{CF}_{r-1,K-N+2}$; $H_1 : \mathcal{CF}_{r-1,K-N+2}(\delta_{c_\theta})$, given b_θ, $\delta^2_{c_\theta} = \dfrac{\text{SNR}\ \lVert \boldsymbol{h}_B \rVert^2\ \sin^2\theta}{1+b_\theta}$.

Bibliography

[1] F. Gini, A. Farina, and M. Greco, "Selected List of References on Radar Signal Processing," *IEEE Transactions on Aerospace and Electronic Systems*, Vol. 37, No. 1, pp. 329-359, January 2001. DOI: 10.1109/7.913696

[2] E. J. Kelly, "An Adaptive Detection Algorithm," *IEEE Transactions on Aerospace and Electronic Systems*, Vol. 22, No. 2, pp. 115-127, March 1986. DOI: 10.1109/TAES.1986.310745

[3] G. De Vito, A. Farina, R. Sanzullo, and L. Timmoneri, "Comparison between two adaptive algorithms for cancellation of noise like and coherent repeater interference," *In Proc. of International Radar Symposium, IRS98*, Munich, Germany, September 1998.

[4] N. B. Pulsone and C. M. Rader, "Adaptive Beamformer Orthogonal Rejection Test," *IEEE Transactions on Signal Processing*, Vol. 49, No. 3, pp. 521-529, March 2001. DOI: 10.1109/78.905870

[5] G. A. Fabrizio, A. Farina, and M. D. Turley, "Spatial Adaptive Subspace Detection in OTH Radar," *IEEE Transactions on Aerospace and Electronic Systems*, Vol. 39, No. 4, pp. 1407-1427, October 2003. DOI: 10.1109/TAES.2003.1261136

[6] F. C. Robey, D. L. Fuhrman, E. J. Kelly, and R. Nitzberg, "A CFAR Adaptive Matched Filter Detector," *IEEE Transactions on Aerospace and Electronic Systems*, Vol. 29, No. 1, pp. 208-216, January 1992. DOI: 10.1109/7.135446

[7] F. Bandiera, O. Besson, and G. Ricci, "An ABORT-like Detector with Improved Mismatched Signals Rejection Capabilities," *IEEE Transactions on Signal Processing*, Vol. 56, No. 1, pp. 14-25, January 2008. DOI: 10.1109/TSP.2007.906690

[8] S. Kraut, L. L. Scharf, and L. T. McWhorter, "Adaptive Subspace Detectors," *IEEE Transactions on Signal Processing*, Vol. 49, No. 1, pp.1-16, January 2001. DOI: 10.1109/78.890324

[9] E. J. Kelly and K. Forsythe, "Adaptive Detection and Parameter Estimation for Multidimensional Signal Models," Lincoln Lab, MIT, Lexington, Tech. Rep. No. 848, April 19, 1989.

[10] S. Z. Kalson, "An Adaptive Array Detector with Mismatched Signal Rejection," *IEEE Transactions on Aerospace and Electronic Systems*, Vol. 28, No. 1, pp. 195-207, January 1992. DOI: 10.1109/7.135445

[11] A. De Maio, "Robust Adaptive Radar Detection in the Presence of Steering Vector Mismatches," *IEEE Transactions on Aerospace and Electronic Systems*, Vol. 41, No. 4, pp. 1322-1337, October 2005. DOI: 10.1109/TAES.2005.1561887

[12] F. Bandiera, A. De Maio, and G. Ricci, "Adaptive CFAR Radar Detection with Conic Rejection," *IEEE Transactions on Signal Processing*, Vol. 55, No. 6, pp. 2533-2541, June 2007. DOI: 10.1109/TSP.2007.893763

[13] F. Bandiera, D. Orlando, and G. Ricci, "A Subspace-based Adaptive Sidelobe Blanker", *IEEE Transactions on Signal Processing*, Vol. 56, No. 9, pp. 4141-4151, September 2008. DOI: 10.1109/TSP.2008.926193

[14] F. Bandiera, O. Besson, D. Orlando, and G. Ricci, "An Improved Adaptive Sidelobe Blanker", *IEEE Transactions on Signal Processing*, Vol. 56, No. 9, pp. 4152-4161, September 2008. DOI: 10.1109/TSP.2008.926191

[15] E. Conte, M. Lops, and G. Ricci, "Asymptotically Optimum Radar Detection in Compound Gaussian Noise," *IEEE Transactions on Aerospace and Electronic Systems*, Vol. 31, No. 2, pp. 617-625, April 1995. DOI: 10.1109/7.381910

[16] S. Kraut and L. L. Scharf, "The CFAR Adaptive Subspace Detector is a Scale-Invariant GLRT," *IEEE Transactions Signal Processing*, Vol. 47, No. 9, pp. 2538-2541, September 1999. DOI: 10.1109/78.782198

[17] C. D. Richmond, "Performance of the Adaptive Sidelobe Blanker Detection Algorithm in Homogeneous Environments," *IEEE Transactions on Signal Processing*, Vol. 48, No. 5, pp. 1235-1247, May 2000. DOI: 10.1109/78.839972

[18] A. De Maio, "Rao Test for Adaptive Detection in Gaussian Interference with Unknown Covariance Matrix," *IEEE Transactions on Signal Processing*, Vol. 55, No. 7, pp. 3577-3584, July 2007. DOI: 10.1109/TSP.2007.894238

[19] H. L. Van Trees, *Optimum Array Processing (Detection, Estimation, and Modulation Theory, Part IV)*, John Wiley & Sons, 2002.

[20] N. R. Goodman, "Statistical Analysis Based on a Certain Multivariate Complex Gaussian Distribution (An Introduction)," *The Annals of Mathematical Statistics*, Vol. 34, No. 1, pp. 152-177, March 1963. DOI: 10.1214/aoms/1177704250

[21] T. L. Grettenberg, "A Representation Theorem for Complex Normal Processes," *IEEE Transactions on Information Theory*, Vol. 11, No. 2, pp. 305-306, April 1965. DOI: 10.1109/TIT.1965.1053774

[22] K. D. Ward, R. J. A. Tough, and S. Watts, *Sea Clutter: Scattering, the K Distribution and Radar Performance,* IET Radar, Sonar and Navigation Series 20, 2008.

[23] M. Sekine and Y. Mao, *Weibull Radar Clutter*, IEE Radar, Sonar, Navigation and Avionics Series 3, 1990.

[24] J. Ward, "Space-Time Adaptive Processing for Airborne Radar," MIT, Lexington, Tech. Rep. No. 1015, December 13, 1994. DOI: 10.1109/22.320768

[25] N. Levanon, *Radar Principles*, John Wiley & Sons, 1988.

[26] H. L. Van Trees, *Detection, Estimation, and Modulation Theory, Part I*, John Wiley & Sons, 2002.

[27] L. L. Scharf, *Statistical Signal Processing: Detection, Estimation, and Time Series Analysis*, Addison-Wesley, 1991.

[28] S. M. Kay, *Fundamentals of Statistical Signal Processing, Vol. 2: Detection Theory*, Prentice-Hall, 1998.

[29] S. M. Kay, "Asymptotically Optimal Detection in Unknown Colored Noise via Autoregressive Modeling," *IEEE Transactions on Acoustics, Speech, and Signal Processing*, Vol. 31, No. 4, pp. 927-940, August 1983. DOI: 10.1109/TASSP.1983.1164156

[30] A. Sheikhi, M.M. Nayebi, and M.R. Aref, "Adaptive detection algorithm for radar signals in autoregressive interference", *IEE Proc. Radar Sonar Navig.*, Vol. 145, No. 5, pp. 309-314, October 1998.

[31] G. Alfano, A. De Maio, and A. Farina, "Model-based adaptive detection of range spread targets," *IEE Proceedings - Radar Sonar Navigation*, Vol. 151, No. 1, pp. 1-10, February 2004. DOI: 10.1049/ip-rsn:20040157

[32] G. T. Capraro, A. Farina, H. Griffiths, and M. C. Wicks, "Knowledge-Based Radar Signal and Data Processing (A Tutorial Review), *IEEE Signal Processing Magazine*, Vol. 23, No. 1, pp. 18-29, January 2006. DOI: 10.1109/MSP.2006.1593334

[33] R. J. Muirhead, *Aspects of Multivariate Statistical Theory*, John Wiley & Sons, 1982. DOI: 10.1002/9780470316559.ch2

[34] K. V. Mardia, J. T. Kent, and J. M. Bibby, *Multivariate Analysis*, Academic Press, 1979.

[35] S. Haykin, *Array Signal Processing*, Prentice-Hall, 1985.

[36] D. Manolakis and G. Shaw, "Detection Algorithms for Hyperspectral Imaging Applications," *IEEE Signal Processing Magazine*, Vol. 19, No. 1, pp. 29-43, January 2002. DOI: 10.1109/79.974724

[37] E. J. Kelly, "Adaptive detection in non-stationary interference-Part III," Massachusetts Institute of Technology, Lincoln Laboratory, Lexington, MA, Tech. Rep. 761, August 1987.

[38] E. J. Kelly, "Performance of an Adaptive Detection Algorithm; Rejection of Unwanted Signals," *IEEE Transactions on Aerospace and Electronic Systems*, Vol. 25, No. 2, pp. 122-133, March 1989.

[39] O. Besson and D. Orlando, "Adaptive Detection in Nonhomogeneous Environments Using the Generalized Eigenrelation," *IEEE Signal Processing Letters*, Vol. 14, No. 10, pp. 731-734, October 2007. DOI: 10.1109/LSP.2007.898355

[40] O. Besson, "Detection in the Presence of Surprise or Undernulled Interference," *IEEE Signal Processing Letters*, Vol. 14, No. 5, pp. 352-354, May 2007. DOI: 10.1109/LSP.2006.888295

[41] C. D. Richmond, "The Theoretical Performance of a Class of Space-Time Adaptive Detection and Training Strategies for Airborne Radar," *In Proc. of 32nd Annual Asilomar Conference on Signals, Systems, and Computers*, Pacific Grove, CA, USA, November 1998. DOI: 10.1109/ACSSC.1998.751541

[42] F. Bandiera, D. Orlando, and G. Ricci, "A Parametric Adaptive Radar Detector," *In Proc. of 2008 IEEE Radar Conference*, Rome, Italy, May 2008.

[43] D. E. Kreithen and A. O. Steinhardt, "Target Detection in Post-Stap Undernulled Clutter," *In Proc. of 29th Annual Asilomar Conference on Signals, Systems, and Computers*, Pacific Grove, CA, USA, November 1995. DOI: 10.1109/ACSSC.1995.540890

[44] C. D. Richmond, "Statistical Performance Analysis of the Adaptive Sidelobe Blanker Detection Algorithm," *In Proc. of 31st Annual Asilomar Conference on Signals, Systems, and Computers*, Pacific Grove, CA, USA, November 1997. DOI: 10.1109/ACSSC.1997.680568

[45] C. D. Richmond, "Performance of a Class of Adaptive Detection Algorithms in Nonhomogeneous Environments," *IEEE Transactions on Signal Processing*, Vol. 48, No. 5, pp. 1248-1262, May 2000. DOI: 10.1109/78.839973

[46] N. B. Pulsone and M. A. Zatman, "A Computationally Efficient Two-Step Implementation of the GLRT," *IEEE Transactions on Signal Processing*, Vol. 48, No. 3, pp. 609-616, March 2000. DOI: 10.1109/78.824657

[47] S. Bose and A. O. Steinhardt, "A maximal invariant framework for adaptive detection with structured and unstructured covariance matrices," *IEEE Transactions on Signal Processing*, vol. 43, No. 9, pp. 2164-2175, September 1995. DOI: 10.1109/78.414779

[48] Y. I. Abramovich, N. K. Spencer, and A. Y. Gorokhov, "Modified GLRT and AMF Framework for Adaptive Detectors," *IEEE Transactions on Aerospace and Electronic Systems*, Vol. 43, No. 3, pp. 1017-1051, July 2007. DOI: 10.1109/TAES.2007.4383590

[49] F. Bandiera, D. Orlando, and G. Ricci, "Adaptive Radar Detection of Distributed Targets under Conic Constraints," *In Proc. of 2008 IEEE Radar Conference*, Rome, Italy, May 2008.

[50] M. Greco, F. Gini, and A. Farina, "Radar Detection and Classification of Jamming Signals Belonging to a Cone Class," *IEEE Transactions on Signal Processing*, Vol. 56, No. 5, pp. 1984-1993, May 2008. DOI: 10.1109/TSP.2007.909326

[51] A. Farina and F. Gini, "Interference Blanking Probabilities for SLB in Correlated Gaussian Clutter Plus Noise," *IEEE Transactions on Signal Processing*, Vol. 48, No. 5, pp. 1481-185, May 2000. DOI: 10.1109/78.839997

[52] H. Lutkepohl, *Handbook of Matrices*, John Wiley & Sons, 1996.

[53] R. A. Horn and C. R. Johnson, *Matrix Analysis*, Cambridge University Press, 1985.

[54] M. Abramowitz and I. A. Stegun, *Handbook of Mathematical Functions with Formulas, Graph, and Mathematical Tables*, Dover Publications Inc., 1972.

[55] C. G. Khatri and C. R. Rao, "Effects of Estimated Noise Covariance Matrix in Optimal Signal Detection," *IEEE Transactions on Acoustic, Speech, and Signal Processing*, Vol. 35, No. 5, pp. 671-679, May 1987. DOI: 10.1109/TASSP.1987.1165185

[56] S. Haykin, *Adaptive Filter Theory*, 3rd edition, Prentice-Hall, 1996.

[57] F. Bandiera, O. Besson, D. Orlando, and G. Ricci, "Theoretical Performance Analysis of the W-ABORT Detector," *IEEE Transactions on Signal Processing*, Vol. 56, No. 5, pp. 2117-2121, May 2008. DOI: 10.1109/TSP.2007.912269